Lecture Notes in Physics

The Lecture Notes in Physics

The series Lecture Notes in Physics (LNP), founded in 1969, reports new developments in physics research and teaching – quickly and informally, but with a high quality and the explicit aim to summarize and communicate current knowledge in an accessible way. Books published in this series are conceived as bridging material between advanced graduate textbooks and the forefront of research and to serve three purposes:

- to be a compact and modern up-to-date source of reference on a well-defined topic

- to serve as an accessible introduction to the field to postgraduate students and nonspecialist researchers from related areas

- to be a source of advanced teaching material for specialized seminars, courses and schools

Both monographs and multi-author volumes will be considered for publication. Edited volumes should, however, consist of a very limited number of contributions only. Proceedings will not be considered for LNP.

Volumes published in LNP are disseminated both in print and in electronic formats, the electronic archive being available at springerlink.com. The series content is indexed, abstracted and referenced by many abstracting and information services, bibliographic networks, subscription agencies, library networks, and consortia.

Proposals should be sent to a member of the Editorial Board, or directly to the managing editor at Springer:

Christian Caron
Springer Heidelberg
Physics Editorial Department I
Tiergartenstrasse 17
69121 Heidelberg / Germany
christian.caron@springer.com

G. Franzese
M. Rubi (Eds.)

Aspects of Physical Biology

Biological Water, Protein Solutions,
Transport and Replication

 Springer

Giancarlo Franzese
Universidad Barcelona
Fac. Fisica
Depto. Fisica Fonamental
Marti i Franquesa, 1
08028 Barcelona
Spain
gfranzese@ub.es

Miguel Rubi
Dept. Fisica Fonamental
Fac. Fisica
Marti i Franquesa, 1
08028 Barcelona
Spain
mrubi@ub.edu

Franzese, G., Rubi, M. (Eds.), *Aspects of Physical Biology: Biological Water, Protein Solutions, Transport and Replication*, Lect. Notes Phys. 752 (Springer, Berlin Heidelberg 2008), DOI 10.1007/978-3-540-78765-5

ISBN: 978-3-642-09757-7   e-ISBN: 978-3-540-78765-5

DOI 10.1007/978-3-540-78765-5

Lecture Notes in Physics ISSN: 0075-8450

Cover design: eStudio Calamar S.L., F. Steinen-Broo, Pau/Girona, Spain

Printed on acid-free paper

9 8 7 6 5 4 3 2 1

springer.com

Preface

The application to Biology of the methodologies developed in Physics is attracting an increasing interest in the scientific community. The physics approach to the study of biological problems has created the new interdisciplinary field of Physical Biology. The aim of this field is to reach a better understanding of the biological mechanisms at the molecular and cellular levels. Statistical Mechanics plays an important role in the development of this new field.

For this reason, we selected as topic and title for the XX Sitges Conference on Statistical Physics "Physical Biology: from Molecular Interactions to Cellular Behavior." As is by now tradition for the Sitges conferences, a number of lectures were subsequently selected, expanded and an updated for publication in the series "Lecture Notes in Physics" to provide both an introduction and an overview to a number of subjects of broader interest and to favor the interchange and cross-fertilization of ideas between biologists and physicists. This volume focuses on three main subtopics: biological water, protein solutions, and transport and replication, presenting for each of them the ongoing debates on the recent results. The role of water in biological processes, the mechanisms of protein folding, the phases and cooperative effects in biological solutions, and the thermodynamic description of replication, transport and neural activity are all subjects that in this volume are revised, based on new experiments and new theoretical interpretations.

The conference itself was held in Sitges (Barcelona, Spain) on 5–9 June, 2006, and was sponsored by several institutions that provided financial support: European Physical Society, Ministerio de Educación y Ciencia of the Spanish Government, Departament d'Universitats, Recerca i Societat de la Informació of the Generalitat de Catalunya, Universitat de Barcelona and the Centre Especial de Recerca (CER) Física de Sistemes Complexos. As in former editions of the conference, the city of Sitges allowed us to use the beautiful Palau Maricel as the lecture hall. We are also very grateful to M. Naspreda, whose contribution as member of the Local Organizing Committee was essential, and to M.-C. Miguel, D. Reguera and J. M. Vilar for their helpful suggestions. Last but not least, we would like to thank all the speakers and participants of the conference, for the high scientific quality of their contributions and for the pleasant atmosphere that they created, and in particular those colleagues who agreed to the effort of providing tutorial accounts of their lectures that make up this exciting volume.

Barcelona,   *Giancarlo Franzese*
December 2007   *Miguel Rubi*

Contents

Part III Transport and Replication

Part I
Biological Water

Dynamics of Water at Low Temperatures and Implications for Biomolecules

P. Kumar, G. Franzese, S.V. Buldyrev and H.E. Stanley

Abstract The biological relevance of water is a puzzle that has attracted much scientific attention. Here we recall what is unusual about water and discuss the possible implications of the unusual properties of water also known as *water anomalies* in biological processes. We find the surprising results that some anomalous properties of water, including results of a recent experiments on hydrated biomolecules, are all consistent with the working hypothesis of the presence of a first-order phase transition between two liquids with different densities at low temperatures and high pressures, which ends in a critical point. To elucidate the relation between dynamic and thermodynamic anomalies, we investigate the presence of this *liquid–liquid* critical point in several models. Using molecular dynamics simulations, we find a correlation between the dynamic transition and the locus of specific heat maxima $C_P^{\max}$ (also known as *Widom line*) emanating from the critical point. We investigate the relation between the dynamic transitions of biomolecules (lysozyme and DNA) and the dynamic and thermodynamic properties of hydration water. We find that the dynamic transition of the macromolecules, sometimes called "protein glass transition" in case of proteins, occurs at the same temperature where the dynamics of hydration water has a crossover and also coincides with the temperature of maximum of specific heat and the maximum of the temperature derivative of the orientational order parameter. Since our simulations are consistent with the possibility that the protein glass transition results from a change in the behavior of hydration water, specifically from crossing the Widom line, we explore in more details the relation

P. Kumar
Center for Studies in Physics and Biology, Rockefeller University, New York, NY 10021, USA, pradeep.kumar@rockefeller.edu

G. Franzese
Departament de Física Fonamental, Universitat de Barcelona, Diagonal 647, Barcelona 08028, Spain, gfranzese@ub.edu

S.V. Buldyrev
Department of Physics, Yeshiva University, 500 West 185th Street, New York, NY 10033, USA, buldyrev@yu.edu

H.E. Stanley
Center for Polymer Studies and Department of Physics, Boston University, Boston, MA 02215, USA, hes@bu.edu

Kumar, P. et al.: *Dynamics of Water at Low Temperatures and Implications for Biomolecules.*
Lect. Notes Phys. **752**, 3–22 (2008)
DOI 10.1007/978-3-540-78765-5_1

between the dynamic crossover and the Widom line in a tractable model for water. We find that the dynamic crossover can be fully explained as a consequence of the thermodynamic and structural changes occurring at the Widom line of water. We, therefore, argue that the so-called "glass transition" of hydrated proteins is just a consequence of the thermodynamic and structural changes of the surrounding water.

1 Introduction

Water is one of the most ubiquitous liquids and perhaps the only one which exists in all the three stable phases (liquid, solid, and gas) at ambient conditions in nature. Although the chemical composition of water is very simple, the physical properties of water make it unique among the other substances. Water has an unusually high boiling and melting point as well as a very high liquid–gas critical temperature compared to the liquids belonging to the same isoelectronic structure as water. Water is also rich in the number of crystal structure it forms at different temperatures and pressures. Indeed, more than thirteen different crystals of water (ices) have been discovered.

Many living beings can survive without water only a few days. This is because water participates in the majority of the biological processes [1], such as the metabolism of nutrients catalysed by enzymes. In order to be effective, the enzymes need to be suspended in a solvent to adopt their active three-dimensional structure. Another important role of water is that it allows the processes of elimination of cellular metabolic residues. It is water through which our cells can communicate and that oxygen and nutrients can be brought to our tissues. Nevertheless, there is no clear reason why water should have an essential role in biological processes.

What makes water more interesting compared to many other liquids are the properties of water, also called *anomalies of water*. Due to its very puzzling nature, water has been the subject of intense studies for decades. One, regarding the well-known anomaly of the density maximum at $4°$ C, dates back to the seventeenth century. It is for this anomaly that ice floats on water and fishes can survive in warm waters below a layer of ice at temperatures well below $0°$ C.

In the following we will discuss the properties of water by, first, describing some of its anomalies, then by presenting a working hypothesis and, next, by testing it. In our discussion we will use different theoretical tools, including computer simulations. The recent dramatic increase of computational power makes it feasible to study the behavior of water at long time scales using computer models, and their quantitative and qualitative agreement with the experiments has been demonstrated by many studies over the last few years.

1.1 Water Anomalies at Low Temperatures

Pure water can been cooled well below the melting point without freezing. Such a water is called *supercooled liquid water*. The supercooled liquid is *metastable* and

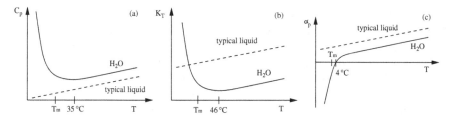

Fig. 1 Schematic representation of different response functions C_P, κ_T, and α_P of liquid water as a function of temperature T. The behavior of a normal liquid is shown as *dashed curves*

the free energy minimum state is the crystal state, i.e. it cannot last forever as a stable liquid phase, but, sooner or later, it will transform into ice.

In the supercooled region most of the anomalous properties of water get more enhanced. In Fig. 1, we show a schematic representation of three different response functions of water, namely the specific heat C_P, the isothermal compressibility κ_T, and the coefficient of thermal expansion α_P, as a function of temperature T at atmospheric pressure. C_P is a measure of enthalpy fluctuation $\langle (\Delta H)^2 \rangle$, hence one would expect that it should decrease as the temperature is decreased; however, for the case of water it increases sharply as the temperature is decreased. Similarly, since κ_T is the measure of volume fluctuations $\langle (\Delta V)^2 \rangle$, it should decrease upon decreasing the temperature. In the case of water, instead, it increases like C_P and seems to diverge at low temperatures in experiments [2]. Figure 1(c) shows the behavior of α_P and since α_P is the measure of cross fluctuations of volume and entropy $\langle \Delta V \Delta S \rangle$, it is positive for normal liquids. However, in the case of water it becomes zero at the temperature of maximum density T_{MD} and negative for $T < T_{\mathrm{MD}}$ at constant pressures, suggesting that below T_{MD} as the volume is increased the entropy decreases. Like other response functions, α_P also seems to diverge at low temperatures in experiments. Dashed curves in Fig. 1 are the schematic representations of the behavior of normal liquids for a comparison. Since the experiments on bulk liquid water cannot be performed below the homogeneous nucleation temperature $T_H \approx -38°$, where the crystal formation is inevitable, it is not possible to test the seeming divergence of response functions at low temperatures. Later we will show, using computer simulations, that indeed the response functions do not diverge but have a maxima at low temperatures.

1.2 Interpretation of the Anomalies

Over the last few decades many experiments and computer simulations of water have been performed and many hypotheses have been put forth to explain these anomalies. Some of the hypotheses were tested and proved wrong, and some of them are still under the scrutiny of careful experiments. There exist two scenarios, namely the singularity-free and liquid–liquid critical point scenario, consistent with the experiments.

In the *singularity-free interpretation* [3, 4] the large increase of response functions seen in the experiments represents only an *apparent* singularity, due to local density fluctuations, with no real divergence. All the anomalies are interpreted as a consequence of the negative volume-entropy cross fluctuations [4].

Another hypothesis is the presence of a phase transition, in the region of deep supercooled water, from high density liquid (HDL) water at high temperature to low density liquid (LDL) water at low temperature. The hypothesized first-order transition line terminates at a critical point known as liquid–liquid critical point of water. This liquid–liquid critical point in water is estimated to be located at around pressure $P \approx 100$ MPa and temperature $T \approx 220$ K [5, 6]. However, this hypothesized critical point lies below the homogeneous nucleation temperature of water and hence no experiments on bulk water can be performed to locate it as water spontaneously freezes.

Although this region cannot be accessed in bulk water, recent experiments on water confined in nanoscale geometries have made it possible to study liquid water in this region, since confinement suppresses the crystallization. Experiments on water confined in the pores of diameter of few nanometers, shed light on the properties of water below the homogeneous nucleation temperature [7, 8]. Specifically, it was found that on decreasing the temperature below $T \approx 225$ K, water shows a dramatic change in the behavior of the dynamics, namely a crossover in the dynamic quantities such as the diffusion coefficient or the correlation times at low T below $T \approx 225$ K.

Although the origin of the crossover has different interpretations [9, 10], the experimental evidences point to indicate that the change in the dynamics is triggered by a local rearrangement of the hydrogen bond network [11].

Using molecular dynamics, in Sect. 2 we provide evidences that this behavior in the change of dynamics of bulk water at low T is consistent with the LL-critical point hypothesis. Next, in Sect. 3 we relate this dynamic crossover for water with the "glass transition" in hydrated proteins suggesting that the changes in the protein dynamics are a consequence of the changes in the behavior of the hydration water. Finally, by using a tractable water model, we show in Sect. 5 that the dynamic crossover can be explained by structural and thermodynamic changes of water.

1.3 Widom Line

By definition, in a first-order phase transitions, thermodynamic state functions such as density ρ and enthalpy H discontinuously change as we cool the system along a path crossing the equilibrium coexistence line [Fig. 2(a), path β]. In a real experiment, this discontinuous change may not occur at the coexistence line since a substance can remain in a supercooled metastable phase until a limit of stability (a spinodal) is reached [12] [Fig. 2(b), path β]. If the system is cooled isobarically along a path above the liquid–gas critical pressure P_c [Fig. 2(b), path α], the state functions continuously change from the values characteristic of a high temperature phase (gas) to those characteristic of a low temperature phase (liquid).

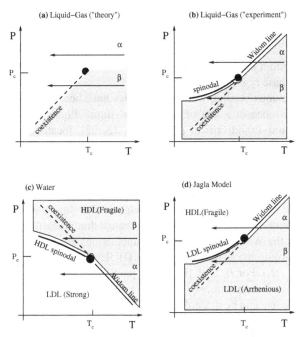

Fig. 2 (**a**) Schematic phase diagram for the critical region associated with a liquid–gas critical point. Shown are the two features displaying mathematical singularities, the critical point (*closed circles*) and the liquid–gas coexistence (*bold dashed curve*). (**b**) Same as (a) with the addition of the gas–liquid spinodal and the Widom line. Along the Widom line, thermodynamic response functions have extrema in their T dependence. The path α denotes a path along which the Widom line is crossed. Path β denotes a path meeting the coexistence line. (**c**) A hypothetical phase diagram for water of possible relevance to the recent neutron scattering experiments by Chen et al. [7, 13] on confined water. The liquid–liquid coexistence, which has a negative sloped coexistence line, generates a Widom line which extends below the critical point, suggesting that water may exhibit a dynamic crossover (non-Arrhenius-to-Arrhenius) transition for $P < P_c$ (path α), while no dynamic changes will occur above the critical point (path β). (**d**) A sketch of the $P - T$ phase diagram for the two-scale Jagla model (see text). For the Jagla potential, as well as for the double-step potential [14] and the shouldered potential [15, 16], the liquid–liquid phase transition line has a positive slope. Upon cooling at constant pressure above the critical point (path α), the liquid changes from a low density state (characterized by a non-glassy Arrhenius dynamics) to a high density state (characterized by glassy Arrhenius dynamics with much larger activation energy) as the path crosses the Widom line. Upon cooling at constant pressure below the critical point (path β), the liquid remains in the LDL phase as long as path β does not cross the LDL spinodal line. Thus, one does not expect any change in the dynamic behavior along the path β, except upon approaching to glass transition at lower T, where one can expect the non-Arrhenius behavior. From [8]

The thermodynamic response functions which are the derivatives of the state functions with respect to temperature [e.g., isobaric heat capacity $C_P \equiv (\partial H/\partial T)_P$] have maxima at temperatures denoted by $T_{\max}(P)$. Remarkably, these maxima are still prominent far above the critical pressure [17, 18, 19, 20, 21], and the values of the response functions at $T_{\max}(P)$ (e.g., $C_P^{\max}$) diverge as the critical point is approached. The lines of the maxima for different response functions asymptotically get closer to one another as the critical point is approached, since all response functions become expressible in terms of the correlation length. This asymptotic line is sometimes called the Widom line and is often regarded as an extension of the coexistence line into the one-phase regime.

As we mentioned in the previous section, water's anomalies have been hypothesized to be related to the existence of a line of a first-order liquid–liquid phase transition terminating at a liquid–liquid critical point [5, 6, 12, 22], located below the homogeneous nucleation line in the deep supercooled region of the phase diagram—sometimes called the "no-man's land" because it is difficult to make direct measurements on the *bulk* liquid phase [6]. In supercooled water, the liquid–liquid coexistence line and the Widom line have negative slopes. Thus, if the system is cooled at constant pressure P_0, computer simulations suggest that for $P_0 < P_c$ [Fig. 2(c), path α] experimentally-measured quantities will change dramatically but continuously in the vicinity of the Widom line (with huge fluctuations as measured by, e.g., C_P) from those resembling the high density liquid (HDL) to those resembling the low density liquid (LDL). For $P_0 > P_c$ [Fig. 2(c), path β], experimentally measured quantities will change discontinuously if the coexistence line is actually seen. However, the coexistence line can be difficult to detect in a pure system due to metastability, and changes will occur only when the spinodal is approached where the HDL phase is no longer stable. The changes in behavior may include not only static quantities like response functions [17, 18, 19, 20, 21] but also dynamic quantities like diffusivity.

In the case of water, a significant change in dynamical properties has been suggested to take place in deeply supercooled states [2, 22, 23, 24, 25, 26]. Unlike other network forming materials [27], water behaves as a non-Arrhenius liquid in the experimentally accessible window [22, 28, 29]. A dynamics is Arrhenius if the relaxation time is well described by

$$\tau = \tau_0 \exp\left[\frac{E_A}{k_B T}\right], \tag{1}$$

where τ_0 is the relaxation time in the large-T limit, k_B is the Boltzmann constant, and E_A is the T-independent activation energy. If E_A depends on T, the dynamics is non-Arrhenius.

Based on analogies with other network forming liquids and with the thermodynamic properties of the amorphous forms of water, it has been suggested that, at ambient pressure, liquid water should show a dynamic crossover from non-Arrhenius behavior at high T to Arrhenius behavior at low T [24, 30, 31, 32, 33, 34]. Using Adam–Gibbs theory [35], the dynamic crossover in water was related to the C_P^{max} line [23, 36]. Also, a dynamic crossover has been associated with the liquid–liquid phase transition in simulations of silicon and silica [37, 38]. Recently a dynamic crossover in confined water was studied experimentally [7, 13, 39, 40] since nucleation can be avoided in confined geometries. Here, we report the recent investigation of experiments on water [7, 13, 40] as arising from the presence of the hypothesized liquid–liquid critical point, which gives rise to a Widom line and an associated dynamics crossover [Fig. 2(c) and 2(d), path α].

2 Bulk Water

Using molecular dynamics simulations [41], Xu et al. [8] studied three models, each of which has a liquid–liquid critical point. Two of the models (the TIP5P [42] and the ST2 [43]) treat water as a multiple site rigid body, interacting via electrostatic site–site interactions complemented by a Lennard-Jones potential. The third model is the spherical "two-scale" Jagla potential with attractive and repulsive ramps which has been studied in the context of liquid–liquid phase transitions and liquid anomalies [21, 31, 32, 33, 44, 45]. For all three models, Xu et al. evaluated the loci of maxima of the relevant response functions, compressibility and specific heat, which coincide close to the critical point and give rise to the Widom line. They found evidence that, for all three potentials, the dynamic crossover occurs just when the Widom line is crossed (Fig. 2).

These findings are consistent with the possibility that the observed dynamic crossover along path α is related to the behavior of C_P, suggesting that enthalpy or entropy fluctuations may have a strong influence on the dynamic properties [21, 23, 38]. Indeed, as the thermodynamic properties change from the high-temperature side of the Widom line to the low-temperature side, $(\partial S/\partial T)_P = C_P/T > 0$ implies that the entropy must decrease. The entropy decrease is most pronounced at the Widom line when $C_P = C_P^{\mathrm{max}}$. Since the configurational part of the entropy, S_{conf}, makes the major contribution to S, we expect that S_{conf} also decreases sharply on crossing the Widom line.

According to Adam–Gibbs theory [35], the diffusivity D changes as

$$D \sim \exp\left[-\frac{A}{T S_{\mathrm{conf}}}\right]. \tag{2}$$

Hence, we expect that D sharply decreases upon cooling at the Widom line. If S_{conf} does not change appreciably with T, then the Adam–Gibbs equation predicts an Arrhenius behavior of D. For both water and the Jagla model, crossing the Widom line is associated with the change in the behavior of the diffusivity. (i) In the case of water, D changes from non-Arrhenius to Arrhenius behavior, while the structural and thermodynamic properties change from those resembling HDL to those resembling LDL, due to the *negative* slope of the Widom line. (ii) For the Jagla potential, D changes from Arrhenius to non-Arrhenius while the structural and thermodynamic properties change from those resembling LDL to those resembling HDL, due to the *positive* slope of the Widom line (Fig. 3).

Thus, these results for bulk water are consistent with the experimental observation in confined water of (i) a dynamic crossover for $P < P_c$ [7, 13], and (ii) a peak in C_P upon cooling water at atmospheric pressure [46], so this work offers a plausible interpretation of the results of [7] as supporting the existence of a hypothesized liquid–liquid critical point.

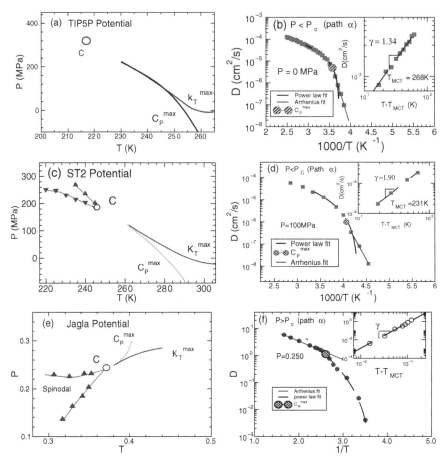

Fig. 3 (**a**) Relevant part of the phase diagram for the TIP5P water potential, showing the liquid–liquid critical point C at $P_c = 320$ MPa and $T_c = 217$ K, the line of maximum of isobaric specific heat C_P^{max} and the line of maximum of isothermal compressibility K_T^{max}. (**b**) D as a function of T for $P = 0$ MPa for ST2 (path α). At high temperatures, D behaves like that of a non-Arrhenius liquid and can be fit by $D \sim (T - T_{MCT})^\gamma$ (also shown in the inset) where $T_{MCT} = 220$ K and $\gamma = 1.942$, while at low temperatures the dynamic behavior changes to that of a liquid where D is Arrhenius. (**c**) The same for the ST2 water potential. The liquid–liquid critical point C is located at $P_c = 246$ MPa and $T_c = 146$ K. (**d**) D as a function of T for $P = 100$ MPa for TIP5P (path α). At high temperatures, D behaves like that of a non-Arrhenius liquid and can be fit by $D \sim (T - T_{MCT})^\gamma$ (also shown in the inset) where $T_{MCT} = 239$ K and $\gamma = 1.57$, while at low temperatures the dynamic behavior changes to that of a liquid where D is Arrhenius. (**e**) Phase diagram for the Jagla potential in the vicinity of the liquid–liquid phase transition. Shown are the liquid–liquid critical point C located at $P_c = 0.24 U_0/a^3$ and $T_c = 0.37 U_0/k_B$ (where U_0 is the attraction energy and a is molecular diameter of the Jagla model), the line of isobaric specific heat maximum C_P^{max}, the line of isothermal compressibility K_T^{max}, and spinodal lines. (**f**) D as a function of T for $P = 0.250 U_0/a^3$ (path α). Along path α, one can see a sharp crossover from the high temperature Arrhenius behavior $D \approx \exp(-1.53/T)$ with lower activation energy to a low temperature Arrhenius behavior $D \approx \exp(-6.3/T)$ with high activation energy, which is a characteristic of the HDL. Adapted from [8]

3 Hydrated Biomolecules

Next we report about the hypothesis [47] that the so called "glass transition" observed in hydrated biomolecules [48, 49, 50, 51, 52, 53, 54, 55, 56, 57, 58, 59, 60, 61, 62] is related to the dynamic crossover described in the previous section. We use quotations for the expression "glass transition" because both terms "glass" and "transition" are used in the literature with an extended meaning. In reality, they refers to a dynamic change observed in hydrated biomolecules. We report here the study by Kumar et al. [47] on the dynamic and thermodynamic behavior of lysozyme and DNA in hydration TIP5P water, performed by means of the molecular dynamics software package GROMACS [63] for (i) an orthorhombic form of hen egg-white lysozyme [64] and (ii) a Dickerson dodecamer DNA [65] at constant pressure $P = 1$ atm, several constant temperatures T, and constant number of water molecules N (NPT ensemble).

The simulation results for the mean square fluctuations $\langle x^2 \rangle$ of both protein and DNA are shown in Fig. 4. The mean square fluctuations $\langle x^2 \rangle$ of the biomolecules is calculated from the equilibrated configurations, first for each atom over 1 ns, and then averaged over the total number of atoms in the biomolecule. Kumar et al. found that $\langle x^2 \rangle$ changes its functional form below $T_p \approx 245$ K, for *both* lysozyme [Fig. 4(a)] and DNA [Fig. 4(b)].

Kumar et al. next calculated C_P by numerical differentiation of the total enthalpy of the system (protein and water) by fitting the simulation data for enthalpy with a fifth order polynomial, and then taking the derivative with respect to T. Figures 5(a) and (b) display maxima of $C_P(T)$ at $T_W \approx 250 \pm 10$ K for both biomolecules.

Further, to describe the quantitative changes in structure of hydration water, Kumar et al. calculated the local tetrahedral order parameter Q [66] for hydration

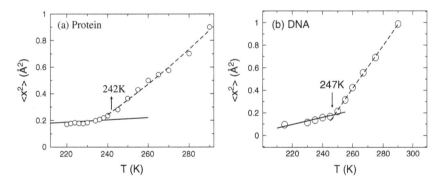

Fig. 4 Mean square fluctuation of (**a**) lysozyme, and (**b**) DNA showing that there is a transition around $T_p \approx 242 \pm 10$ K for lysozyme and around $T_p \approx 247 \pm 10$ K for DNA. For very low T one would expect a linear increase of $\langle x^2 \rangle$ with T, as a consequence of harmonic approximation for the motion of residues. At high T, the motion becomes non-harmonic and is fitted by a polynomial. The dynamic crossover temperature T_p is determined from the crossing of the linear fit for low T and the polynomial fit for high T. The error bars is determined by changing the number of data points in the two fitting ranges. From [47]

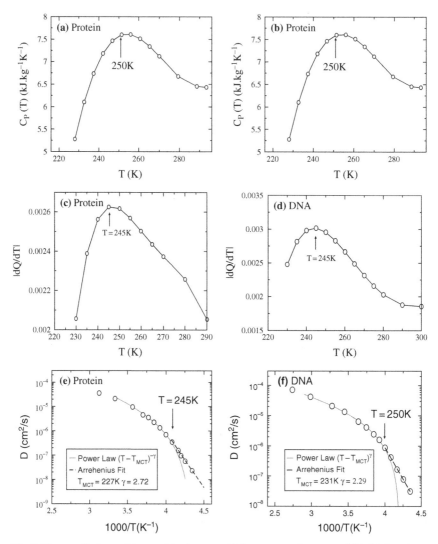

Fig. 5 The specific heat of the combined system (**a**) lysozyme and water, and (**b**) DNA and water, display maxima at 250 ± 10 K and 250 ± 12 K respectively, which are coincident within the error bars with the temperature T_p where the crossover in the behavior of $\langle x^2 \rangle$ is observed in Fig. 4. Derivative with respect to temperature of the local tetrahedral order parameter Q for (**c**) lysozyme and (**d**) DNA hydration water. A maximum in $|dQ/dT|$ at Widom line temperature suggests that the rate of change of local tetrahedrality of hydration water has a maximum at the Widom line. Diffusion constant of hydration water surrounding (**e**) lysozyme and (**f**) DNA shows a dynamic transition from a power law behavior to an Arrhenius behavior at $T_\times \approx 245 \pm 10$ K for lysozyme and $T_\times \approx 250 \pm 10$ K for DNA, around the same temperatures where the behavior of $\langle x^2 \rangle$ has a crossover, and C_P and $|dQ/dT|$ have maxima. From [47]

water surrounding lysozyme and DNA. Figures 5(c) and (d) show that the rate of increase of Q has a maximum at 245 ± 10 K for lysozyme and DNA hydration water respectively; the same temperatures of the crossover in the behavior of mean square fluctuations.

Upon cooling, the diffusivity of hydration water exhibits a dynamic crossover from non-Arrhenius to Arrhenius behavior at the crossover temperature $T_\times \approx 245 \pm 10$ K for lysozyme [Fig. 5(e)] and $T_\times \approx 250 \pm 10$ K for DNA [Fig. 5(f)]. The coincidence of $T_\times$ with T_p within the error bars indicates that the behavior of the protein and the DNA is strongly coupled with the behavior of the surrounding solvent, in agreement with recent experiments on hydrated protein and DNA [48] which found the crossover in side-chain fluctuations at $T_p \approx 225$ K [48]. Note that $T_\times$ is much higher than the glass transition temperature, estimated for TIP5P as $T_g = 215$ K [67]. Thus, this crossover is not likely to be related to the glass transition in water.

The fact that $T_p \approx T_\times \approx T_W$ is evidence of the correlation between the changes in protein fluctuations [Fig. 4(a)] and the hydration water thermodynamics [Fig. 5(a)]. Thus, these results are consistent with the possibility that the protein "glass transition" is related to the Widom line. In the next section, we will explore with more details the relation between the thermodynamics, structural properties, and dynamics in the vicinity of the Widom line. Here we observe that crossing the Widom line corresponds to a continuous but rapid transition of the properties of water from those resembling the properties of a local HDL structure for $T > T_W(P)$ to those resembling the properties of a local LDL structure for $T < T_W(P)$ [7, 8, 47]. A consequence is the expectation that the fluctuations of the protein residues in predominantly LDL-like water (more ordered and more rigid) just below the Widom line should be smaller than the fluctuations in predominantly HDL-like water (less ordered and less rigid) just above the Widom line.

The quantitative agreement of the results for both DNA and lysozyme (Figs. 4 and 5) suggests that it is indeed the changes in the properties of hydration water that are responsible for the changes in dynamics of the protein and DNA biomolecules. All the results suggest that these changes occur at the Widom line.

4 Other Evidences of Changes at the Widom Line

In the previous section, we concentrated on reviewing the evidence for changes in dynamic transport properties, such as diffusion constant and relaxation time upon crossing the Widom line. In addition, a series of other phenomena has been observed in computer simulations and experiments around the Widom line:

- A sharp breakdown of Stokes–Einstein relation and a data collapse if T is replaced by $T - T_W$ [68, 69, 70].
- The appearance of a "boson peak" on cooling below the Widom line [68].
- The occurrence of a new transport mechanism below the Widom line, showing "dynamic heterogeneities" [68, 69, 70].

- A dynamic crossover in the relaxation time for Q fluctuations below the Widom line [71].
- Structural changes in $g(r)$ and its Fourier transform $S(q)$ [72].
- A sharp drop in T derivative of zero-frequency structure factor.

It is possible that all these phenomena that appear on crossing the Widom line are in fact not coincidences, but are related to the changes in local structure occurring when the system goes from the HDL-like side to the LDL-like side of the Widom line. To understand better this possibility, we report an investigation [73] of a tractable model, in which the simulations can be complemented with analytic calculations. In [73], indeed, it has been found an explicit relation between the thermodynamics and the dynamics that allows to calculate the crossover directly from thermodynamics quantities. Since the crossover appears to be related only to the properties of water, the authors focus on the behavior of bulk water to reduce the complexity of the analysis.

5 Relation Between Thermodynamics and Dynamics

In [73], Kumar et al. proposed and verified a relation between dynamics and thermodynamics that shows how the crossovers described in previous sections are a direct consequence of the structural change at the temperature of the specific heat maximum $T(C_P^{\mathrm{max}})$. They adopted a tractable model for water [74, 75, 76, 77] and expressed the relevant free energy barrier for the local rearrangement of the molecules in terms of the probability p_B of forming bonds. By mean field calculations and Monte Carlo simulations they found that the variation of p_B is the largest at $T(C_P^{\mathrm{max}})$. If the phase diagram displays a liquid–liquid critical point, $T(C_P^{\mathrm{max}})$ coincides with $T_W(P)$, the Widom line. On approaching the maximum pressure P_W^{max} along $T_W(P)$, the crossover changes from sharp, as in water at ambient P [7], to smooth, a result that is explained in terms of the P-dependence of the free energy barrier.

We consider a cell model that reproduces the fluid phase diagram of water and other tetrahedral network forming liquids [74, 75, 76, 77]. For the sake of clarity, we focus on water to explain the motivation of the model. The model is based on the experimental observations that on decreasing P at constant T, or on decreasing T at constant P, (i) water displays an increasing local tetrahedrality [78, 79], (ii) the volume per molecule increases at sufficiently low P or T, and (iii) the O-O-O angular correlation increases [80], consistent with simulations [81, 82].

The system is divided into cells $i \in [1, \ldots, N]$ on a regular square lattice, each containing a molecule with volume $v \equiv V/N$, where $V \geq Nv_0$ is the total volume of the system, and v_0 is the hard-core volume of one molecule. The cell volume v is a continuous variable that gives the mean distance $r \equiv v^{1/d}$ between molecules in d dimensions. The van der Waals attraction between the molecules is represented by a truncated Lennard-Jones potential with characteristic energy $\epsilon > 0$

$$U(r) \equiv \begin{cases} \infty & \text{for } r \leq R_0 \\ \epsilon \left[\left(\frac{R_0}{r} \right)^{12} - \left(\frac{R_0}{r} \right)^6 \right] & \text{for } r > R_0 \, , \end{cases} \tag{3}$$

where $R_0 \equiv v_0^{1/d}$ is the hard-core distance [74, 75, 76, 77].

Each molecule i has four bond indices $\sigma_{ij} \in [1, \ldots, q]$, corresponding to the nearest-neighbor cells j. When two nearest-neighbor molecules have the facing σ_{ij} and σ_{ji} in the same relative orientation, they decrease the energy by a constant J, with $0 < J < \epsilon$, and form a (non-bifurcated) hydrogen bond. The choice $J < \epsilon$ guarantees that bonds are formed only in the liquid phase. The bond interaction is accounted for by a term in the Hamiltonian

$$\mathscr{H}_B \equiv -J \sum_{\langle i,j \rangle} \delta_{\sigma_{ij}\sigma_{ji}}, \tag{4}$$

where the sum is over nearest-neighbor cells, and $\delta_{a,b} = 1$ if $a = b$ and $\delta_{a,b} = 0$ otherwise. For water at high P and T a more dense, collapsed and distorted, local structure with bifurcated hydrogen bonds is consistent with the experiments. Bifurcated hydrogen bonds decrease the strength of the network and favor the hydrogen bond breaking and re-formation. The model simplifies the situation by assuming that (a) only non-bifurcated, i.e. normal, hydrogen bond decrease the energy of the system and (b) the local density changes as function of the number of normal hydrogen bonds, consistent with the observation [80] that at low P and T there is a better separation between the first neighbors and the second neighbors, favoring normal hydrogen bonds and the tetrahedral order.

The density decrease for $T < T_{MD}(P)$, the temperature of maximum density, is represented by an average increase of the molar volume due to a more structured network. If the total volume increases by an amount $v_B > 0$ for each bond formed [4, 74, 75, 76, 77], then the molar volume is

$$v = v' + 2p_B v_B, \tag{5}$$

where v' is the molar volume without taking into account the bond and p_B is the probability of forming a bond. The increase of the angular molecular correlation is modeled by introducing an intramolecular (IM) interaction of energy $0 < J_\sigma < J$,

$$\mathscr{H}_{IM} \equiv -J_\sigma \sum_i \sum_{(k,\ell)_i} \delta_{\sigma_{ik}\sigma_{i\ell}}, \tag{6}$$

where $\sum_{(k,\ell)_i}$ denotes the sum over the bond indices of the molecule i.

The total energy of the system is the sum of (3), (4), and (6). They perform mean field calculations and Monte Carlo simulations in the NPT ensemble [74, 75, 76, 77] for a system with $J/\epsilon = 0.5$, $J_\sigma/\epsilon = 0.05$, $v_B/v_0 = 0.5$, $q = 6$. The study of two square lattices with 900 and 3600 cells shows no appreciable size effects. Below the T_{MD} line, in the supercooled region, the model displays a first-order phase transition between a LDL at low P and T and a HDL at high P and T along a line terminating

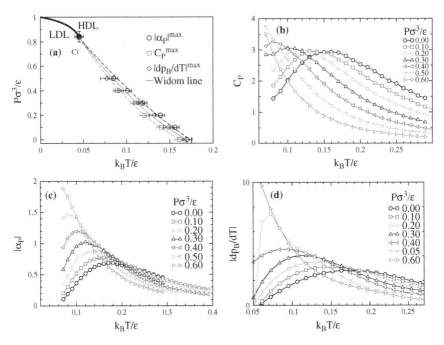

Fig. 6 (a) Phase diagram below T_{MD} line shows that $|dp_B/dT|^{\max}$ ($\Diamond$) coincides with the Widom line $T_W(P)$ (*solid line*) within error bars: C' is the HDL–LDL critical point, end of first-order phase transition line (*thick line*) [74, 75, 76, 77]; symbols are maxima for $N = 3600$ of $|\alpha_P|^{\max}$ ($\bigcirc$), $C_P^{\max}$ ($\Box$), and $|dp_B/dT|^{\max}$ ($\Diamond$); *upper* and *lower dashed line* are quadratic fits of $|\alpha_P|^{\max}$ and $C_P^{\max}$, respectively, consistent with C'; $|\alpha_P|^{\max}$ and $C_P^{\max}$ are consistent within error bars. Maxima are estimated from panels (**b**), (**c**) and (**d**), where each quantity is shown as functions of T for different $P < P_{C'}$. In (**d**) $|dp_B/dT|^{\max}$ is the numerical derivative of p_B from simulations. From [73]

in the liquid–liquid critical point C' [74, 75, 76, 77] [Fig. 6(a)]. For $J_\sigma \to 0$, C' moves toward $T = 0$ [83], suggesting that the *singularity free* scenario is a limiting case of the liquid–liquid critical point scenario [83]. For $J_\sigma > 0$, $P_W^{\max}$ coincides with $P_{C'}$, while for $J_\sigma = 0$ it is a point on the $T = 0$ axis [83].

Kumar et al. found that different response functions such as C_P, α_P [Figs. 6(b) and (c)] show maxima and these maxima increase and seem to diverge as the critical pressure is approached, consistent with the picture of Widom line that we discussed for other water models in the sections above. Moreover, the temperature derivative of the probability of forming hydrogen bonds dp_B/dT displays a maximum in the same region where the other thermodynamic response functions have maxima, suggesting that the fluctuations in the p_B is maximum at the Widom line temperature $T_W(P)$ [Fig. 6(d)].

To further test if this model system also displays a dynamic crossover as found in the other models of water, the total spin relaxation time of the system as a function of T for different pressures is studied. The results show that the liquid is more non-Arrhenius for increasing P [Fig. 7(a)]. For $P = 0$ [Fig. 7(b)] by decreasing T there

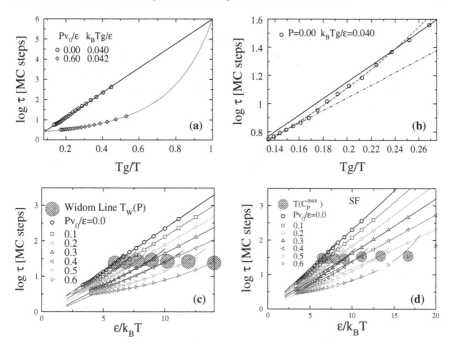

Fig. 7 (a) Angell plot of logarithm of relaxation time τ as a function of T_g/T shows a large difference between $Pv_0/\epsilon = 0$ and 0.6; T_g is defined in such a way that $\tau(T_g) = 10^6$ MC steps $\sim 10^7$ ps [26]; for $Pv_0/\epsilon = 0.6$ ($\diamond$) the τ can be described as a VFT function (7) (*curved line*), while for $P = 0$ ($\bigcirc$) as an Arrhenius function (*straight line*). Note that the data for $Pv_0/\epsilon = 0.6$ extends only down to temperatures where the system has non-Arrhenius dynamics. This suggests that the system becomes increasingly non-Arrhenius as the pressure is increased. (b) A detailed observation at $P = 0$ shows three regimes: Arrhenius at low T_g/T with low activation energy (*slope of dash-dot line*), non-Arrhenius (VFT, *dash line*) at intermediate T, and again Arrhenius at high T_g/T with higher activation energy (*slope of solid line*) as in experiments on water. (c) At $0 \leq Pv_0/\epsilon \leq 0.4$ the Arrhenius crossover occurs at T consistent with $T_W(P)$ (large ⊛); the error on $T_W(P)$ is approximately equal to symbol size; thick and thin lines represent VFT and Arrhenius fits. (d) Dynamic crossover for the singularity free (SF) scenario, with crossover temperature at $T(C_P^{max})$. *Solid* and *dashed lines* represent Arrhenius and non-Arrhenius fits, respectively. Notice that the dynamic crossover occurs at approximately the same value of τ for all seven values of pressure studied, as in panel (c)

is a crossover from Arrhenius to the Vogel–Fulcher–Tamman (VFT) function

$$\tau^{VFT} = \tau_0^{VFT} \exp\left[\frac{T_1}{T - T_0}\right], \tag{7}$$

where τ_0^{VFT}, T_1, and T_0 are all fitting parameters. At lower temperatures there is another crossover from VFT back to Arrhenius. The Arrhenius activation energy at low T is higher than that at high T, consistent with experiments at ambient P for both bulk water [30, 84] and confined water [7, 39].

As the pressure is increased toward the critical pressure P_C of the system, this dynamic crossover in τ at low T gets smoother [Fig. 7(c)], suggesting that the dynamics of water will be more non-Arrhenius as the critical point is approached.

For all P the crossover occurs at $T_W(P)$ within the error bars [Fig. 7(c)], confirming the idea proposed on the base of simulations of detailed models for water reported in the previous sections [8, 47]. We observe that the low-T behavior is characterized by an activation energy—the slope in Fig. 7(c)—that decreases for increasing P, as in experiments for confined water [7], and that the time needed to reach the maximum correlation length is almost independent of the position along $T_W(P)$, being $\log \tau(T_W) \simeq 1.5$MC steps $\simeq 15$ ps [26] for any P. For completeness we study the system also in the case of *singularity free scenario*, corresponding to $J_\sigma = 0$. For $J_\sigma = 0$ the crossover is at $T(C_P^{\max})$, the temperature of $C_P^{\max}$ [Fig. 7(d)]. Kumar et al. next calculated the Arrhenius activation energy $E_A(P)$ from the low-T slope of $\log \tau$ vs. $1/T$ [Fig. 8(a)]. They extrapolated the temperature $T_A(P)$ at which τ reached a fixed macroscopic time $\tau_A \geq \tau_C$. We choose $\tau_A = 10^{14}$ MC steps > 100 [26] [Fig. 8(b)]. We find that $E_A(P)$ and $T_A(P)$ decrease upon increasing P in both scenarios, providing no distinction between the two interpretations. Instead, we found a dramatic difference in the P dependence of the quantity $E_A/(k_B T_A)$ in the two scenarios, increasing for the liquid–liquid critical point scenario and approximately constant for the singularity free scenario [Fig. 8(c)].

To better understand their findings, Kumar et al. [73] developed an expression for τ in terms of thermodynamic quantities, which allows to explicitly calculate $E_A/(k_B T_A)$ for both scenarios. For any activated process, in which the relaxation from an initial state to a final state passes through an excited transition state, $\ln(\tau/\tau_0) = \Delta(U + PV - TS)/(k_B T)$, where $\Delta(U + PV - TS)$ is the difference in free energy between the transition state and the initial state. Consistent with results from simulations and experiments [85, 86], Kumar et al. proposed that at low T the mechanism to relax from a less structured state (lower tetrahedral order) to a more structured state (higher tetrahedral order) corresponds to the breaking of a bond and the simultaneous molecular reorientation for the formation of a new bond. The transition state is represented by the molecule with a broken bond and more tetrahedral IM order. Hence,

$$\Delta(U + PV - TS) = J p_B - J_\sigma p_{IM} - P v_B - T \Delta S, \tag{8}$$

where p_B and p_{IM}, the probability of a satisfied IM interaction, can be directly calculated. To estimate ΔS, the increase of entropy due to the breaking of a bond, they use the mean field expression $\Delta S = k_B[\ln(2N p_B) - \ln(1 + 2N(1 - p_B))]\bar{p}_B$, where $\bar{p}_B$ is the average value of p_B above and below $T_W(P)$.

They next tested that the expression of $\ln(\tau/\tau_0)$, in terms of ΔS and (8),

$$\ln \frac{\tau}{\tau_0} = \frac{J p_B - J_\sigma p_{IM} - P v_B}{k_B T} - \bar{p}_B \ln \frac{2N p_B}{1 + 2N(1 - p_B)} \tag{9}$$

describes the simulations well, with minor corrections at high T. Here $\tau_0 \equiv \tau_0(P)$ is a free fitting parameter equal to the relaxation time for $T \to \infty$. From (9) they found

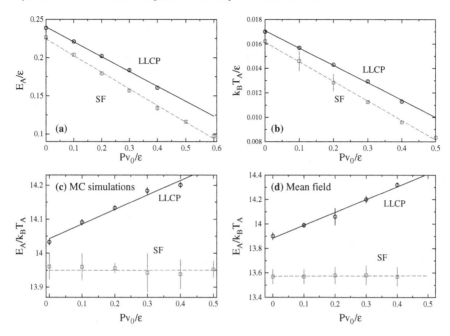

Fig. 8 Effect of pressure on the activation energy E_A. (**a**) Demonstration that E_A decreases linearly for increasing P for both the *liquid–liquid critical point* (LLCP) and the *singularity free* (SF) scenarios. The *lines* are linear fits to the simulation results (*symbols*). (**b**) T_A, defined such that $\tau(T_A) = 10^{14}$ MC steps > 100 [26] decreases linearly with P for both scenarios. (**c**) P dependence of the quantity $E_A/(k_B T_A)$ is different in the two scenarios. In the LLCP scenario, $E_A/(k_B T_A)$ increases with increasing P, and it is approximately constant in the SF scenario. The *lines* are guides to the eyes. (**d**) Demonstration that the same behavior is found using the mean field approximation. In all the panels, where not shown, the error bars are smaller than the symbol sizes

that the ratio $E_A/(k_B T_A)$ calculated at low T increases with P for $J_\sigma/\epsilon = 0.05$, while it is constant for $J_\sigma = 0$, as from their simulations [Fig. 8(d)].

6 Conclusions

Recent experiments at very low temperatures, below the crystal homogeneous nucleation temperature of bulk water, have revealed the relation between the dynamics of water and that of proteins or DNA [47, 48]. The experiments on hydration water as well as on confined water [7, 39, 40, 87], and the corresponding simulations [8, 47], find a change in the water relaxation dynamics that suggests a crossover from a non-Arrhenius dynamics at higher T to Arrhenius dynamics at very low T [8], as predicted for bulk water [30].

The simulations show that the dynamics becomes Arrhenius when the liquid is cooled isobarically below the temperature of maximum specific heat $T(C_P^{\max})$ and that water is more tetrahedral for $T < T(C_P^{\max})$ [47]. In one possible interpretation of the anomalous properties of water, $T(C_P^{\max})$ coincides with the Widom line emanating from a hypothesized liquid–liquid critical point C' [6], the terminus of a liquid–liquid phase transition line that separates a low-density liquid and a high-density liquid.

Based on the study of the probability p_B of forming a Hydrogen bond, we have seen that it is possible to show [73] how the crossover at $T(C_P^{\max})$ is a consequence of a local relaxation process associated with breaking a bond and reorienting the molecule. The investigation reported here allow to find the relation between dynamics and thermodynamics in terms of the relevant activation barrier [73]. The crossover at $T(C_P^{\max})$ is regulated by (9), as a consequence of a local breaking of a bond and simultaneous molecular reorientation to increase the structural order. This is consistent with what is observed in SPC/E water [85] and with recent experimental conclusions [9]. Kumar et al. [73] has found the expression for the relevant free energy barrier [Eq. (8)] and how it depends on pressure. The barrier rapidly increases with $1/T$ above $T(C_P^{\max})$, while has only a weak dependence on T below $T(C_P^{\max})$, giving rise to the crossover.

In conclusion, the investigations reported here show that the dynamics of water at low temperature is a consequence of the structural changes of the network formed by the hydrogen bonds. Our results suggest that the change in the dynamic behavior of water around the line of maxima of specific heat could be the cause of the so-called "glass transition" in hydrated proteins or DNA. Hence the dynamics of biomolecules turns out to be dominated by the changes in the hydration water.

Acknowledgements We thank C. A. Angell, W. Kob, L. Liu, F. Mallamace, P. H. Poole, S. Sastry, and NSF for financial support. G. F. also thanks the Spanish Ministerio de Educación y Ciencia (Programa Ramón y Cajal and Grant No. FIS2004-03454).

References

1. F. Franks: *Biochemistry and Biophysics at Low Temperatures* (Cambridge University Press, Cambridge 1985)
2. C. A. Angell: Annu. Rev. Phys. Chem. **55**, 559 (2004)
3. H. E. Stanley and J. Teixeira: J. Chem. Phys. **73**, 3404 (1980)
4. S. Sastry, P. G. Debenedetti, F. Sciortino, and H. E. Stanley: Phys. Rev. E **53**, 6144 (1996)
5. P. H. Poole, F. Sciortino, U. Essmann, and H. E. Stanley: Nature (London) **360**, 324 (1992)
6. O. Mishima and H. E. Stanley: Nature (London) **392**, 164 (1998)
7. L. Liu, S.-H. Chen, A. Faraone, C.-W. Yen, and C.-Y. Mou: Phys. Rev. Lett. **95**, 117802 (2005)
8. L. Xu, P. Kumar, S. V. Buldyrev, S.-H. Chen, P. Poole, F. Sciortino, and H. E. Stanley: Proc. Natl. Acad. Sci. U.S.A. **102**, 16558 (2005)
9. J. Swenson: Phys. Rev. Lett. **97**, 189801 (2006)
10. S. Cerveny, J. Colmenero, and A. Alegria: Phys. Rev. Lett. **97**, 189802 (2006)
11. S.-H. Chen, L. Liu, and A. Faraone: Phys. Rev. Lett. **97**, 189803 (2006)
12. P. G. Debenedetti and H. E. Stanley: Phys. Today **56**, 40 (2003)
13. A. Faraone, L. Liu, C. Y. Mou, C. W. Yen, and S. H. Chen: J. Chem. Phys. **121**, 10843 (2004)
14. G. Franzese, G. Malescio, A. Skibinsky, S. V. Buldyrev, and H. E. Stanley: Nature (London) **409**, 692 (2001)

15. G. Franzese: Differences between discontinuous and continuous soft-core attractive potentials: the appearance of density anomaly (preprint). J. Mol. Phys. (2007)
16. A. B. de Oliveira, G. Franzese, P. A. Netz, and M. C. Barbosa: Water-like hierarchy of anomalies in a continuous spherical shouldered potential. J. Chem. Phys. **128**, 064901 (2008), URL http://arxiv.org/abs/0706.2838
17. M. A. Anisimov, J. V. Sengers, and J. M. H. Levelt Sengers: In: *Aqueous System at Elevated Temperatures and Pressures: Physical Chemistry in Water, Steam and Hydrothermal Solutions*, ed. by D. A. Palmer, R. Fernandez-Prini, and A. H. Harvey (Elsevier, Amsterdam 2004) pp. 29–71
18. J. M. H. Levelt: Measurements of the compressibility of argon in the gaseous and liquid phase. Ph.D. thesis, University of Amsterdam, Van Gorkum and Co., Assen (1958)
19. A. Michels, J. M. H. Levelt, and G. Wolkers: Physica **24**, 769 (1958)
20. A. Michels, J. M. H. Levelt, and W. de Graaff: Physica **24**, 659 (1958)
21. L. Xu, S. Buldyrev, C. A. Angell, and H. E. Stanley: Phys. Rev. E **74**, 031108 (2006)
22. P. Debenedetti: J. Phys.: Condens. Matter **15**, R1669 (2003)
23. F. W. Starr, C. A. Angell, and H. E. Stanley: Physica A **323**, 51 (2003)
24. C. A. Angell: J. Phys. Chem. **97**, 6339 (1993)
25. P. Kumar, S. V. Buldyrev, F. W. Starr, N. Giovambattista, and H. E. Stanley: Phys. Rev. E **72**, 051503 (2005)
26. P. Kumar, G. Franzese, S. V. Buldyrev, and H. E. Stanley: Phys. Rev. E **73**, 041505 (2006)
27. J. Horbach and W. Kob: Phys. Rev. B **60**, 3169 (1999)
28. E. W. Lang and H.-D. Lüdemann: Angew. Chem. Int. Ed. Engl. **21**, 315 (2004)
29. F. X. Prielmeier, E. W. Lang, R. J. Speedy, and H.-D. Lüdemann: Phys. Rev. Lett. **59**, 1128 (1987)
30. K. Ito, C. T. Moynihan, and C. A. Angell: Nature (London) **398**, 492 (1999)
31. E. A. Jagla: J. Chem. Phys. **111**, 8980 (1999)
32. E. A. Jagla: J. Phys.: Condens. Matter **11**, 10251 (1999)
33. E. A. Jagla: Phys. Rev. E **63**, 061509 (2001)
34. H. Tanaka: J. Phys. Condens. Matter **15**, L703 (2003)
35. G. Adam and G. H. Gibbs: J. Chem. Phys. **43**, 139 (1965)
36. P. H. Poole, I. Saika-Voivod, and F. Sciortino: J. Phys.: Condens. Matter **17**, L431 (2005)
37. S. Sastry and C. A. Angell: Nat. Matter. **2**, 739 (2003)
38. I. Saika-Voivod, P. H. Poole, and F. Sciortino: Nature (London) **412**, 514 (2001)
39. R. Bergman and J. Swenson: Nature (London) **403**, 283 (2000)
40. F. Mallamace, M. Broccio, C. Corsaro, A. Faraone, U. Wanderlingh, L. Liu, C. Y. Mou, and S. H. Chen: J. Chem. Phys. **124**, 161102 (2006)
41. D. C. Rapaport: *The Art of Molecular Dynamics Simulation* (Cambridge University Press, Cambridge 1995)
42. M. W. Mahoney and W. L. Jorgensen: J. Chem. Phys. **112**, 8910 (2000)
43. F. H. Stillinger and A. Rahman: J. Chem. Phys. **57**, 1281 (1972)
44. P. Kumar, S. V. Buldyrev, F. Sciortino, E. Zaccarelli, and H. E. Stanley: Phys. Rev. E **72**, 021501 (2005)
45. L. M. Xu, I. Ehrenberg, S. V. Buldyrev, and H. E. Stanley: J. Phys.: Condens. Matter **18**, S2239 (2006)
46. S. Maruyama, K. Wakabayashi, and M. Oguni: In: *Slow Dynamics in Complex Systems: Third International Symposium on Slow Dynamics in Complex Systems, AIP Conf. Proc. No. 708*, ed. by M. Tokuyama and I. Oppenheim (AIP, New York 2004) pp. 675–676
47. P. Kumar, Z. Yan, L. Xu, M. G. Mazza, S. V. Buldyrev, S.-H. Chen, S. Sastry, and H. E. Stanley: Phys. Rev. Lett. **97**, 177802 (2006)
48. S. H. Chen, L. Liu, E. Fratini, P. Baglioni, A. Faraone, and E. Mamontov: Proc. Natl. Acad. Sci. U.S.A. **103**, 9012 (2006)
49. J.-M. Zanotti, M.-C. Bellissent-Funel, and J. Parello: Biophys. J. **76**, 2390 (1999)
50. D. Ringe and G. A. Petsko: Biophys. Chem. **105**, 667 (2003)
51. J. M. Wang, P. Cieplak, and P. A. Kollman: J. Comp. Chem. **21**, 1049 (2000)

52. E. J. Sorin and V. S. Pande: Biophys. J. **88**, 2472 (2005)
53. B. F. Rasmussen, A. M. Stock, D. Ringe, and G. A. Petsko: Nature (London) **357**, 423 (1992)
54. D. Vitkup, D. Ringe, G. A. Petsko, and M. Karplus: Nat. Struct. Biol. **7**, 34 (2000)
55. A. P. Sokolov, H. Grimm, and R. Kahn: J. Chem. Phys. **110**, 7053 (1999)
56. W. Doster, S. Cusack, and W. Petry: Nature (London) **337**, 754 (1989)
57. J. Norberg and L. Nilsson: Proc. Natl. Acad. Sci. U.S.A. **93**, 10173 (1996)
58. M. Tarek and D. J. Tobias: Phys. Rev. Lett. **88**, 138101 (2002)
59. M. Tarek and D. J. Tobias: Biophys. J. **79**, 3244 (2000)
60. H. Hartmann, F. Parak, W. Steigemann, G. A. Petsko, D. R. Ponzi, and H. Frauenfelder: Proc. Natl. Acad. Sci. U.S.A. **79**, 4967 (1982)
61. A. L. Tournier, J. Xu, and J. C. Smith: Biophys. J. **85**, 1871 (2003)
62. A. L. Lee and A. J. Wand: Nature (London) **411**, 501 (2001)
63. E. Lindahl, B. Hess, and D. van der Spoel: J. Mol. Mod. **7**, 306 (2001)
64. P. J. Artymiuk, C. C. F. Blake, D. W. Rice, and K. S. Wilson: Acta Crystallogr. B **38**, 778 (1982)
65. H. R. Drew, R. M. Wing, T. Takano, C. Broka, S. Tanaka, K. Itakura, and R. E. Dickerson: Proc. Natl. Acad. Sci. U.S.A. **78**, 2179 (1981)
66. J. R. Errington and P. D. Debenedetti: Nature (London) **409**, 318 (2001)
67. I. Brovchenko, A. Geiger, and A. Oleinikova: J. Chem. Phys. **123**, 044515 (2005)
68. S.-H. Chen, F. Mallamace, C.-Y. Mou, M. Broccio, C. Corsaro, A. Faraone, and L. Liu: Proc. Natl. Acad. Sci. U.S.A. **103**, 12974 (2006)
69. P. Kumar: Proc. Natl. Acad. Sci. U.S.A. **103**, 12955 (2006)
70. P. Kumar, S. V. Buldyrev, S. R. Becker, P. H. Poole, F. W. Starr, and H. E. Stanley: Proc. Natl. Acad. Sci. U.S.A. **104**, 9575 (2007)
71. P. Kumar, S. V. Buldyrev, and H. E. Stanley: Space-time correlations in the orientational order parameter and the orientational entropy of water (preprint). (2007)
72. P. Kumar, S. V. Buldyrev, and H. E. Stanley: In: *Soft Matter under Extreme Pressures: Fundamentals and Emerging Technologies*, ed. by S. J. Rzoska and V. Mazur, Proc. NATO ARW, Odessa, Oct. 2005 (Springer, Berlin Heidelberg New York 2006)
73. P. Kumar, G. Franzese, and H. E. Stanley: Phys. Rev. Lett. **100**, 105701 (2008)
74. G. Franzese and H. E. Stanley: J. Phys.: Condens. Matter **14**, 2201 (2002)
75. G. Franzese and H. E. Stanley: Physica A **314**, 508 (2002)
76. G. Franzese, M. I. Marques, and H. E. Stanley: Phys. Rev. E **67**, 011103 (2003)
77. G. Franzese and H. E. Stanley: J. Phys.: Condens. Matter **19**, 205126 (2007)
78. G. Darrigo, G. Maisano, F. Mallamace, P. Migliardo, and F. Wanderlingh: J. Chem. Phys. **75**, 4264 (1981)
79. Angell, C. A. and Rodgers, V.: J. Chem. Phys. **80**, 6245 (1984)
80. A. K. Soper and M. A. Ricci: Phys. Rev. Lett. **84**, 2881 (2000), and references cited therein
81. E. Schwegler, G. Galli, and F. Gygi: Phys. Rev. Lett. **84**, 2429 (2000)
82. P. Raiteri, A. Laio, and M. Parrinello: Phys. Rev. Lett. **93**, 087801 (2004), and references cited therein
83. K. Stokeley, M. G. Mazza, G. Franzese, and H. E. Stanley: A general model for the thermal behavior of supercooled water, submitted URL arXiv:0805.3468v1<http://arxiv.org/abs/0805.3468v1>
84. I. Kohl, L. Bachmann, A. Hallbrucker, E. Mayer, and T. Loerting: Phys. Chem. Chem. Phys. **7**, 3210 (2005)
85. Laage, D. and Hynes, J. T.: Science **311**, 832 (2006)
86. Tokmakoff, A.: Science **317**, 54 (2007)
87. M. A. Ricci, F. Bruni, P. Gallo, M. Rovere, and A. K. Soper: J. Phys.: Condens. Matter **12**, A345 (2000)

Anomalous Behaviour of Supercooled Water and Its Implication for Protein Dynamics

J. Swenson, H. Jansson and R. Bergman

Abstract Water is the foundation of life, and without it life as we know it would not exist. An organism consists to a large extent of water and, apart from a few larger reservoirs, almost all water in a living organism is closely associated with surfaces of biomolecules of different kinds. This so-called biological water is known to affect the dynamics of biomaterials such as proteins, which in turn is crucial for its functions. However, how and why the surrounding environment affects the dynamics of proteins and other biomolecules is still not fully understood. Recently, it was suggested [Fenimore et al. PNAS 2004, **101** 14408] that local and more global protein motions are slaved (or driven) by the local β-relaxation and the more large-scale cooperative α-relaxation in the surrounding solvent, respectively. In this chapter we present results from dielectric measurements on myoglobin in water-glycerol mixtures that support this slaving idea. Moreover, we show how confined supercooled water changes its dynamical behaviour from a low temperature Arrhenius behaviour to a high temperature non-Arrhenius behaviour at a certain temperature (around 200 K), and then we discuss likely explanations for the crossover and its consequence for protein dynamics.

1 Introduction

Water is the only chemical compound on earth that occurs naturally in the solid, liquid and vapour phases [1]. Water exhibits many unique properties on which life, as far as we know, is dependent. For instance, water's polarity, high dielectric constant

J. Swenson
Department of Applied Physics, Chalmers University of Technology, SE-412 96, Göteborg, Sweden, f5xjs@fy.chalmers.se

H. Jansson
Department of Applied Physics, Chalmers University of Technology, SE-412 96, Göteborg, Sweden, helen.jansson@fy.chalmers.se

R. Bergman
Department of Applied Physics, Chalmers University of Technology, SE-412 96, Göteborg, Sweden, f5xrb@fy.chalmers.se

Swenson, J. et al.: *Anomalous Behaviour of Supercooled Water and Its Implication for Protein Dynamics.* Lect. Notes Phys. **752**, 23–42 (2008)
DOI 10.1007/978-3-540-78765-5_2

and small molecular size makes it an excellent solvent for polar and ionic compounds [2, 3]. It has unique hydration properties for biological macromolecules,
where it, for example, is involved in the folding of proteins and, consequently their
function, by exclusion of non-polar compounds, i.e. the hydrophobic effect [2].
This effect is important in biological systems where it, for instance, is involved
in the stabilisation of the folded protein structure as well as in the formation of cell
membranes [2, 4]. The water in an organism is furthermore involved in the temperature regulation and resistance against dehydration [5]. Moreover, water has unusual high melting and boiling points as well as unusual high viscosity and surface
tension compared to similar low-molecular-weight compounds [2]. Many of these
properties, among others, are due to the hydrogen bonds between adjacent water
molecules [2, 4]. One water molecule can participate in four hydrogen bonds, which
results in a network formation. The formation of hydrogen bonds between adjacent
water molecules results in a microscopic structure of bulk water that is currently
a matter of debate. The accepted picture of the structure has until recently been
that the water molecules tend to form a tetrahedral hydrogen bonded structure [6],
although these hydrogen bonds have very short lifetimes (in order of picoseconds
at room temperature [7]). However, recently this structural picture has been questioned [8], and it has instead been proposed that the water molecules are connected
in chains or rings rather than in a complete network. The structure of bulk water
is, thus, still not established and further investigations are required. When water, as
any other liquid, is cooled, its viscosity increases and if crystallisation is avoided the
liquid will be supercooled below its melting temperature. The supercooled state is a
disordered state where the liquid behaves as a meta-stable, and viscous, liquid. As
the supercooled liquid is further cooled, the viscosity increases rapidly and, when
the viscosity exceeds a value of 10^{13} poise ($= 10^{12}$ Nsm^{-2}), the supercooled liquid
is frozen on experimental time-scales and behaves like a solid. Such an amorphous
solid is called a glass [9]. The transition from a meta-stable supercooled liquid to
an amorphous solid is called the glass transition [9]. One important characteristic of
supercooled liquids and the glass transition is the main structural relaxation process,
the so-called α-relaxation, which occurs in the supercooled liquid and is due to collective rearrangements [9, 10]. The configurational changes in the system that give
rise to this relaxation process are directly coupled to the macroscopic viscosity [9],
and they become dramatically slower with decreasing temperature. At the glass transition temperature, T_g, the relaxation time (τ) of the α-process has reached a value
of about 100 s (in analogy with the 10^{13} poise for viscosity). In addition to the
main relaxation many glass-forming liquids also show weaker secondary relaxation
processes, where the most common are denoted as β-relaxations. β-relaxations are
faster and of more local nature than the α-relaxation and persists below T_g [11], see
Fig. 1. At a temperature above the glass transition temperature, the two relaxation
processes merge and only a single relaxation process is observed at higher T [9, 10].

The α- and β-relaxation times generally show different temperature behaviours.
The fast secondary β-processes normally exhibits an Arrhenius temperature dependence (1) whereas the increase in relaxation time (or viscosity) with decreasing temperature of the α-relaxation generally follows a Vogel-Fulcher-Tammann behaviour
(2) [12, 13, 14]

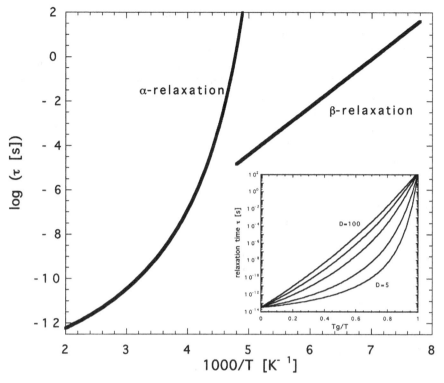

Fig. 1 Arrhenius plot showing typical temperature dependences of the fast β-relaxation and the slower α-relaxation for a glass forming liquid. The α- and β-relaxation times generally show different temperature behaviours, where the β-relaxation normally exhibits an Arrhenius temperature dependence (1) whereas the increase in relaxation time (or viscosity) with decreasing temperature of the α-relaxation generally follows a VFT behaviour (2). The inset shows the characteristic VFT dependences of the α-relaxation for strong and fragile liquids. The curvature (i.e. deviation from Arrhenius behaviour) determines the fragility and is given by the parameter D in (2). The larger curvature (smaller value of D) the more fragile is the liquid. In the figure, *curves* for $D = 5, 10, 25,$ 50 and 100 are shown

$$\tau = \tau_0 \exp(E_a/k_B T) \tag{1}$$

$$\tau = \tau_0 \exp(DT_0/T - T_0) \tag{2}$$

where τ_0 is the microscopic relaxation time extrapolated to infinite temperature, which usually corresponds to quasi-lattice and molecular vibrations of the order of 10^{-14} s. E_a is the activation energy, T_0 is the temperature where τ goes to infinity and D describes the degree of non-Arrhenius behaviour. The so-called strong glass-formers, according to Angell' strong-fragile classification scheme [15, 16], have an α-relaxation time (or viscosity) with a temperature dependence close to Arrhenius behaviour (a large D value in the VFT function), while fragile glass-formers show a much more rapid increase of the α-relaxation time (or viscosity) close to T_g (i.e. a low value of D) [15, 16], see inset in Fig. 1.

The glass transition temperature T_g of bulk water is still widely debated and currently under active research. A $T_g = 136$ K has been accepted for many years [17, 18], but is now questioned and it has recently been suggested to be located around 165 K [19, 20, 21]. The problem with this temperature is that it is located in the so-called *No man's land*, which is the temperature region between 150 and 235 K where supercooled bulk water immediately crystallises [6] and, hence, studies of bulk water are impossible. One way to overcome this problem is to study water in confinements. In a confined environment the water molecules are affected by surfaces, and water layers with different properties are formed [22, 23]. This will change the orientation of adjacent molecules in a way that depends on the chemical nature of the surface (i.e. whether the surface is hydrophilic or hydrophobic, or positively or negatively charged) [24] (and references therein). This orientation will in turn affect the interaction between the water molecules and, as a result, the probability of forming the network structure necessary for crystallisation [25, 26, 27, 28]. In addition, the motion of the water molecules in a confined geometry is normally restricted due to the interaction with the surfaces, which results in a slower dynamics compared to bulk water [24, 28, 29]. Besides that confinements make studies of pure water easier it is, furthermore, of central importance for the understanding of how the water dynamics is affected by surfaces of different kinds, which in turn is relevant for studies of water close to biomolecular surfaces, the so-called biological water, see Fig. 2.

Many studies have been performed on the dynamics of water in different types of confinements. Several of these studies have shown that the temperature dependence of the relaxation time for the observed main process changes at a certain temperature (typically around 200 K), see e.g. [30, 31, 32], from a low temperature Arrhenius behaviour to a Vogel–Fulcher–Tammann behaviour at higher temperatures, i.e. (1) and (2), respectively. The crossover in temperature dependence has been explained by that the water undergoes a fragile-to-strong transition [33]. Such a transition has earlier been proposed to explain the facts that water is a very fragile liquid close to its melting point, and that the relaxation times do not extrapolate to the previously suggested T_g at 136 K. However, there are indications that the observed transitions for confined water are not due to a true fragile-to-strong transition [34], which will be discussed below.

The water content of a living organism is around 70% of the total weight [3], and a large fraction of this water is closely associated (within 5 Å [35]) with proteins and other biomolecules. Water in biological systems is very important since

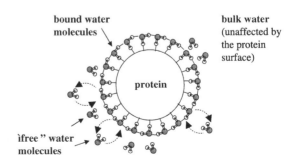

Fig. 2 The water surrounding a biological molecule is called biological water. This water can be divided into three different layers dependent on the distance to, and interaction with the biomolecular surface

folding, structure, stability, mobility and function of biomolecules (such as proteins and nucleic acids) and biological assemblies (such as membranes) are dependent on the aqueous environment [1]. The water surrounding a protein molecule is generally divided into three hydration layers dependending on the distance to and consequently the interaction with the protein surface [36], see Fig. 2. The water molecules closest to the protein surface make up the first hydration layer. These molecules are tightly bound via hydrogen bonds to the protein surface. The water molecules in the second hydration layer, which are less affected by the protein surface, interact with the first hydration layer. The third hydration layer is the surrounding bulk-like water. The first and the second hydration layer constitute the so-called hydration shell. The motion of the water molecules in the hydration shell is, as for water at other surfaces, affected by the interaction with the surface and the dynamics is, thus, restricted and slower compared to the dynamics of bulk water [37, 38, 39, 40]. This retarded motion is however of great biological importance since it enables protein motions and proton diffusion along the protein surface [35], which both are necessary for biological processes. Translational diffusion of surrounding water molecules promotes the changes of the protein structure that is necessary for protein function [41, 42]. By restraining the translational motions of the water the protein motions are reduced in a similar way as in the dehydrated state [41]. The level of hydration is, thus, another important factor for protein function. It is for instance known that the critical point for percolation of protons on the surface is close to the hydration level for enzymatic activity [43], and that totally dry proteins show no biological function [44]. The protein function starts when the hydration level is around 0.2 g water per gram of protein [44], and with a further increase in hydration the protein dynamics, and consequently also the function is increased. For full biological function of the protein an equal amount of water and protein is needed [43]. The reason for the increase in function with increasing water content can be explained by the plasticizing effect of water on proteins, i.e. increase in flexibility of the protein that is necessary for protein function [45] (and references therein). But still it is far from fully understood how the environment surrounding proteins affect their function.

Recently, it has been suggested that the hydration shell and the surrounding bulk water promote different kinds of protein motions and functions [46]. It is suggested that large-scale conformational changes in the protein are coupled to collective motions (i.e. the α-relaxation) in the surrounding bulk water and are, hence, not present for dehydrated proteins. This is in contrast to the more local β-like motions in protein, which seems to be coupled to the more local dynamics in the hydration shell, see Fig. 3. However, this does not mean that a solvent relaxation and its associated protein motion have to appear on the same time-scale, i.e. having the same relaxation rate, but that the relaxation times of the processes show similar temperature dependences (i.e. similar activation energies at a given temperature), as shown in Fig. 3 [46]. Thus, in general, protein motions and the related solvent motions occur on different time-scales, and the reason for this is the complex energy landscape of proteins, see Fig. 4. The protein conformational energy landscape consists of a large number of energy valleys, i.e. conformational substates, separated by energy barriers, and each relaxation or motion of the protein involve a large number of small steps between substates, each given by a specific rate that in turn depends on specific solvent motions as described above.

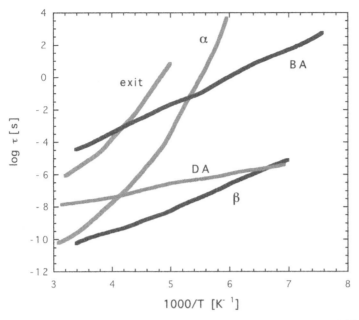

Fig. 3 The relaxation time for the process *exit*, which shows the escape of CO from Myoglobin, follows the same temperature dependence as the α-relaxation of the solvent (3:1 glycerol:water mixture), whereas the migration of CO within the protein molecule (process *BA*) shows the same temperature behaviour as a secondary relaxation of water in the hydration shell, here denoted as β. Slightly modified version of figure from Fenimore et al. [46], reproduced with permission

2 Apparent Fragile to Strong Transitions of Confined Supercooled Water

The relaxation behaviour of deeply supercooled liquids is in general described by the viscosity-related main (α) relaxation and one or several secondary (β) relaxation processes, where the main relaxation process, as described above, generally shows some degree of non-Arrhenius behaviour, whereas the β-processes normally are described by the Arrhenius equation. In Fig. 5a we show temperature dependences of structural relaxation processes measured by dielectric spectroscopy and quasielastic neutron scattering (QENS), where a "normal" liquid behaviour is represented by results for propylene glycol (PG). From this figure it is evident that the temperature dependence of the α-relaxation for both bulk PG and PG confined to one molecular layer in the interplatelet space of a Na-vermiculite clay can be described by the same VFT equation, i.e. there is no obvious dynamic crossover and the confinement has no or very little effect on the relaxational behaviour. This is in contrast to the behaviour of the main relaxation observed for water in widely different confinements. As shown in Fig. 5a for water confined to two molecular layers in the same type of Na-vermiculite clay, as well as for water confined in 10 Å pores of a molecular sieve and for hydration water on hemoglobin, an anomalous behaviour

free energy

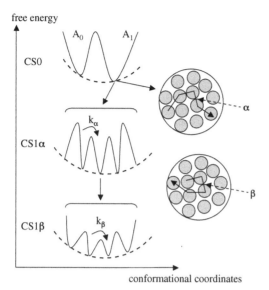

conformational coordinates

Fig. 4 Schematic 1D and 2D drawings of the complex protein energy landscape to the *left* and *right*, respectively (redrawn after [46]). The figure shows that the protein can assume a large number of different conformations, so-called conformational substates, where each substate is separated from the others by energy barriers. On the top level, denoted by CS0, the protein can adopt different conformations that involve different functions. In each CS0 (called A_0 and A_1) there are a large number of substates (CS1α), which all involve the same function but with different rates (k), and contains numerous additional substates (CS1β). The α- and the β-fluctuations of the solvent determine transitions between CS1α and CS1β substates of the protein, respectively

is observed. In these systems, as well as in other systems of interfacial water, the main relaxation time of the water is strongly affected by interactions with the host material and, more importantly, a crossover in the dynamical behaviour from a low temperature Arrhenius behaviour to a VFT temperature dependence at higher temperatures occurs somewhere in the temperature interval of 180–250 K depending on the size and geometry of the confined water and the chemical nature of the host material.

Recently, it was suggested that this type of crossover in temperature dependence for the water dynamics in confinements is due to a so-called fragile-to-strong transition [31], as mentioned above. However, there are indications that the observed dynamic crossover is not due to a true fragile-to-strong transition, [34]. The reason for this is that there are no clear evidences for that the main relaxation process in the deeply supercooled regime (i.e. below the crossover temperature) corresponds to the cooperative α-relaxation. Thus, since the fragility of a liquid refers to the temperature dependence of its viscosity or α-relaxation time, the fragility cannot even be defined if the α-relaxation is not observed. Instead, the dielectric data as well as other experimental results suggest that this low temperature process is due to a secondary and more local β-relaxation. First, the dielectric loss peak has a symmetric shape in the frequency domain and the associated relaxation time exhibits

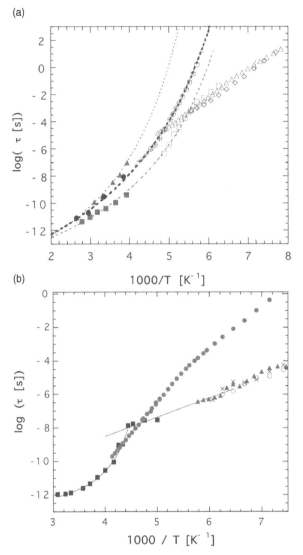

Fig. 5 (**a**) Average relaxation times, obtained by QENS (*filled symbols*) and dielectric spectroscopy (*open symbols*), for water in a clay [30, 49] (*triangles*), in 10 Å pores of a molecular sieve [50, 51] (*squares*) and for water in hydrated hemoglobin [52] (*diamonds*). For comparison, average relaxation times for bulk PG (*crosses*) and for PG confined in clay [53, 54] (*circles*) are shown to represent a normal liquid behaviour. The high temperature dependences of the relaxation times have been described (and extrapolated) by the VFT equation. (**b**) Relaxation times for the main process of water confined in approximately 20 Å pores of MCM-41, obtained by QENS [31] (*squares*) and dielectric spectroscopy [55] (*filled circles*). In addition, the fastest universal dielectric relaxation process of supercooled water in clay [56] (*triangles*), phospholipids membrane DMPC [57] (*open circles*) and myoglobin [58] (*crosses*) are shown for comparison. The *lines* are fits to the high and low temperature QENS data, taken from [31]. Figure 5b is taken from [58]

an Arrhenius temperature dependence. Both features are typical for β-relaxations. Second, as shown in Fig. 6, no calorimetric glass transition can be clearly observed for such confined water. One may argue that this is a natural result due to the severe confinement of the water, but the fact is that for other liquids confined in the same or similar host materials there is no difficulty to observe their calorimetric glass transitions, even for more severe confinements than in the present cases [47]. Thus, the absence of any calorimetric T_g for the confined supercooled water suggests that no dielectric α-relaxation is present in the same temperature regime. Third, in [48] a similar relaxation process was observed for rapidly quenched bulk water by electron spin resonance measurements, even in the so-called *No man's land* (150–235 K) where bulk water is mainly crystalline. This fact further supports that the observed relaxation process is of local character (in this case the relaxation of water molecules in the interface between different crystalline regions). Fourth, further support is that the relaxation process is of local character and it is so weakly dependent on the nature of the interactions with the host material [21], in contrast to the merged α–β relaxation above the crossover temperature, see Fig. 5a. Fifth, the temperature for which the relaxation time of this process reaches 100 s is around 130 K, which means that if it is due to the α-relaxation a T_g is expected at about this temperature. However, since the α-relaxation at high temperatures is slower in confinement than in bulk (and the difference appears to increase with decreasing temperature), it seems unphysical to have a T_g around 130 K for the confined water. Therefore, a likely explanation for the apparent transition in the dielectric data is that it is caused by a merging of the observed β-relaxation with a non-observable α-relaxation at a temperature around 200 K [34]. However, this interpretation only explains the apparent fragile-to-strong transition in dielectric data of confined supercooled water. In the cases of QENS and ^{1}H NMR diffusion measurements on confined and hydration water [31, 32, 59, 60, 61], the even more pronounced crossover in the observed dynamics at about 225 K is more likely due to a decoupling of a water diffusion process from the structural relaxation of the water molecules, see Fig. 5b. This interpretation of the apparent fragile-to-strong transition in the QENS and ^{1}H NMR data is the most probable since the observed low-temperature Arrhenius behaviour extrapolates to the fastest dielectric process of deeply supercooled water [56]. This fast dielectric process is likely due to local reorientations of single water molecules, which are directly related to the long range diffusion in water, as observed by ^{1}H NMR diffusion measurements [61]. Generally, the diffusion constant D is related to the structural relaxation time through the Stokes–Einstein relation but this is obviously not the case for confined supercooled water. The reason for this might be found by investigating the diffusion of water molecules in ice [62, 63], which is almost as fast as in confined supercooled water [61] despite that no true structural relaxation process exists in ice. Hence, the long-range diffusion of water molecules, as measured by these techniques, becomes decoupled from the structural relaxation of the water molecules at $T \sim 225$ K, and as a result an apparent fragile-to-strong transition is observed in experimental data probing the diffusion.

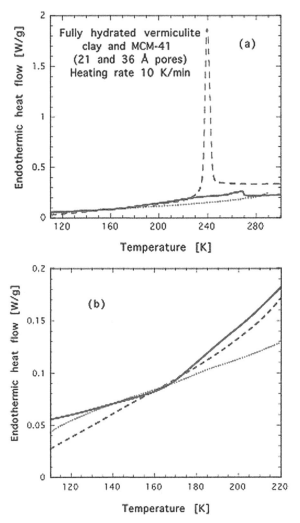

Fig. 6 (**a**) DSC data of fully hydrated Na-vermiculite clay (*dotted line*) and MCM-41 with pore diameters of 21 Å (*solid line*) and 36 Å (*dashed line*). In (**b**) the figure is plotted on an expanded scale to verify that no calorimetric T_g can be observed for any of the samples. The figure is taken from [58]

3 Relation Between Solvent and Protein Dynamics

It is, as described above, well known that proteins cannot function without surrounding hydration water. The reason for this is that protein functions involve conformational changes of the proteins, which, in some way, are promoted by the solvent. However, exactly how the solvent enables protein fluctuations is still a matter of discussion [41, 46, 64, 65, 66]. Since function and dynamics of proteins are closely related to the dynamics of the surrounding water it is of fundamental importance to investigate not only the protein dynamics but also the dynamics of the water molecules associated to protein surfaces. For systematic investigations of the relation between protein and solvent dynamics, it is furthermore useful to change both the total solvent content as well as the dynamical properties of the solvent. Therefore, in a recent study we have investigated myoglobin in different water–glycerol mixtures of various water content. The advantage of using this type of mixture is fourfold: first, the proteins are biologically active even for high glycerol contents (up to about 90 wt% glycerol [H. Frauenfelder, Private Communication 2006]); second, the addition of glycerol makes it more difficult for the water to crystallise in the supercooled regime; third, the glycerol molecules are forming hydrogen bonds to each other and to biological surfaces (just as water molecules do); and fourth, the dynamical properties of the solvent can be substantially changed by changing the water–glycerol ratio. In Fig. 7, we show a typical three-dimensional representation of the dielectric loss, i.e. the imaginary part of the dielectric function versus frequency ($10^{-2} - 10^6$ Hz) and temperature (120–350 K) for myoglobin in a water–glycerol mixture of 50 wt% water at the solvent level $h = 2$ (i.e. 2 g solvent per gram

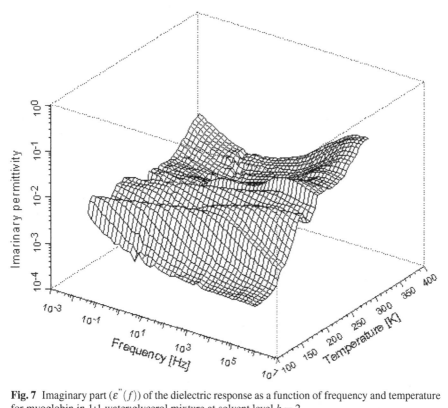

Fig. 7 Imaginary part ($\varepsilon''(f)$) of the dielectric response as a function of frequency and temperature for myoglobin in 1:1 water:glycerol mixture at solvent level $h = 2$

of protein). From this figure it is clear that the sample exhibits a complicated relaxation behaviour with several relaxation processes. In Fig. 8 the relaxation times for these processes are shown together with relaxation times obtained for myoglobin in water–glycerol mixtures of 20 and 33 wt% water, respectively. For each sample two solvent processes are observed, where the fastest one is likely due to local reorientations of single water molecules as described in the previous section. The second fastest, which is the dominating process at the lowest temperatures, is due to the main relaxation of the solvent, and it can be seen that it becomes significantly faster with increasing water fraction for a given total solvent content ($h = 2$). For water fractions of less than 50 wt% this relaxation process corresponds to the α-relaxation. However, in the case of 50 wt% water, as seen in Fig. 8, there is a similar crossover (around 200 K) from a high temperature non-Arrhenius behaviour to a low temperature Arrhenius dependence as shown in Fig. 5a for confined water. This crossover is, as discussed above, a result of a merging of a dielectrically non-observable α-relaxation with the β-relaxation [34]. Thus, below the crossover temperature this process corresponds to the more local β-relaxation of the water. The remaining relaxation processes, which are observed at higher temperatures, are more difficult to determine due to overlap with a Maxwell–Wagner interfacial polarisation process [67]. This Maxwell–Wagner process is due to that the electrodes were covered with a 100 μm thick teflon film in order to suppress an otherwise dominating contribution to the spectra from conductivity and electrode polarisation at high temperatures. Thus, although Fig. 7 shows that there is no significant low frequency dispersion due to conductivity, the used teflon film had the drawback to cause interfacial polarisation in the sample–teflon interface. Since this Maxwell–Wagner polarisation gives rise to a dielectric loss peak in the same frequency range as the protein dynamics appear, it is difficult to accurately extract the processes originating from protein dynamics. In the near future we will overcome this problem by using a considerably thinner teflon film, which will result in a shift of the Maxwell–Wagner process to a lower frequency where no protein processes are present. However, despite this difficulty an attempt is here made to extract the protein process from the dielectric raw data. Depending on the glycerol/water ratio we are able to extract 2–4 different types of motions, although the exact origins of all of them are presently not clear. In the case of the glycerol-rich solvents (i.e. 20 and 33 wt% water), where no vanishing of the cooperative α-relaxation occurs, also the protein processes seem to show a cooperative behaviour (i.e. a non-Arrhenius temperature dependence) at all temperatures. The origin of the fastest one of these cooperative protein processes (in the case of 20 wt% water) is still unclear, but the second fastest process, which is the fastest protein process observed for the sample with 33 wt% water, has been attributed to motions of polar side groups [52, 68]. The slower processes are most likely due to more global conformational fluctuations in the protein [69] or proton displacements along hydrogen bonded solvent molecules adsorbed on the protein surface [70].

Also worth noting in Fig. 8 is, furthermore, that the protein seems to undergo a similar dynamic crossover as the solvent does in the case of the most water-rich solvent. Below the crossover temperature around 200 K only one protein process

Fig. 8 Temperature dependences of the relaxation times for the observed dielectric processes obtained for myoglobin in different water–glycerol mixtures. The solvent level is for all three samples $h = 2$ and the water fractions in the solvent are 20, 33 and 50 wt%, as given in the figures. The symbols correspond most likely to fast local motions of single water molecules (*crosses*), the dominating relaxation process of the solvent (*filled circles*), motions of polar side groups (*filled squares*) and conformational protein fluctuations (*filled triangles*). The origin of the two latter processes as well as the remaining processes (denoted by *open squares*) are not fully clear (see text), but in the sample with 50 wt% water the fastest of them (observed at low temperatures) should represent some kind of local non-cooperative motions in the protein

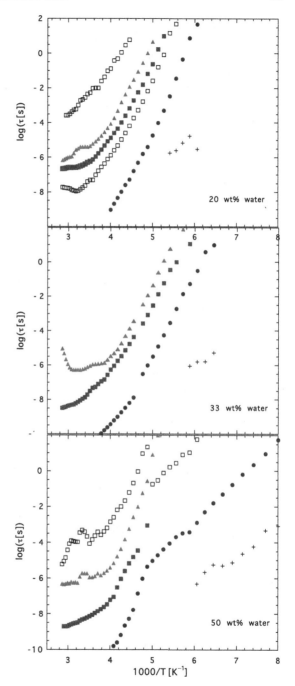

is observed. It also shows an Arrhenius temperature dependence, and is attributed to local non-cooperative protein motions. At the crossover temperature, where we believe the local β-relaxation in the solvent merges with the non-observable α-relaxation, there is a significant change in the protein dynamics. Cooperative protein processes appear, suggesting that the onset of more large-scale solvent motions give rise to a glass-transition-like phenomenon in the protein. It should here be noted that this glass-transition-like behaviour observed with dielectric spectroscopy has nothing to do with the "dynamic transition" (or sometimes even the denomination "glass transition" is used) commonly observed with QENS [71, 72, 73] and Mössbauer spectroscopy [74, 75, 76]. With these techniques considerably faster dynamics is probed than the time scale of about 100 s for the viscosity-related α-relaxation that is commonly used to define T_g. Since the experimental time scale probed in, for instance, QENS is often less than a nanosecond it is obvious that glass-transition-like phenomena cannot be probed with this technique. In such studies the observed "dynamic transition" is nothing else than an onset of anharmonic motions on the experimental time-scale (i.e. a relaxation process has moved into the experimental time-window so it can be observed). This also means that the temperature for which the "dynamical transition" occurs depend on the experimental energy resolution (i.e. the shortest times that can be probed by the instrument).

In order to elucidate the relation between protein and solvent dynamics in more detail, we show in Fig. 9 the relaxation times of the fastest protein processes as a function of the relaxation time for the main process in the solvent. From this figure it is evident that there are linear dependences (i.e. slopes of unity in the log-log plots) for all the shown protein processes and sample compositions (which also has been observed for myoglobin in pure water [34]). Thus, provided that the protein processes have been correctly extracted, as discussed above, we observe the expected behaviour for solvent-slaved protein processes [46, 64]. Furthermore, as is evident for the sample with 50 wt% water, it appears as the local non-cooperative protein process at low temperatures (below the crossover temperature) has the same activation energy as the low temperature β-relaxation in the solvent, suggesting that this protein process is determined by the β-relaxation in the solvent. At higher temperatures (above the crossover temperature) and higher glycerol contents only cooperative protein processes can be observed and these exhibit a similar non-Arrhenius temperature dependence as the solvent does in the same temperature range, indicating that these protein processes are determined by the α-relaxation in the solvent. Therefore, our findings support a recent study by Fenimore et al. [46], where it was proposed that the most local protein motions are slaved by the β-relaxation in the hydration shell whereas the more global conformational changes of a protein are slaved by the α-relaxation in the bulk solvent, although our results suggest that cooperative large-scale protein and solvent processes are present even for relatively low solvent contents where no true bulk solvent is present.

Since large-scale conformational protein fluctuations are required for full biological activity of proteins [75] an important consequence of the present findings is that proteins must be almost inactive in water rich solvents at low temperatures where no α-relaxation is present in the solvent. Thus, for instance, cryopreservation

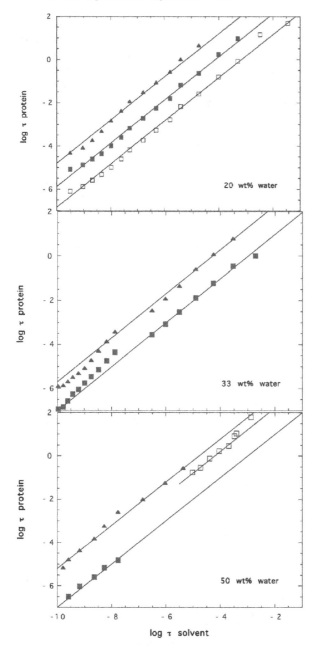

Fig. 9 The relaxation times for the protein processes (symbols as in Fig. 8) are shown as a function of the relaxation time for the main process in the solvent. Provided that these protein processes have been correctly extracted, it is evident that there is an almost linear dependence for all protein processes and samples, as expected for solvent-slaved protein motions

of protein-rich food does not only slow down the biologically most important protein processes, it even prevents them from occurring if the temperature is reduced below the crossover temperature. The behaviour is in close analogy with normal glass-forming liquids for which the time scale of the cooperative α-relaxation basically goes to infinity and only local β-relaxations remain below the glass transition temperature. In fact, it has been shown that hydrated proteins even show a calorimetric glass-transition-like feature in the same temperature range as we observe the dynamic crossover [77]. However, as mentioned in the introduction, the proteins do not undergo this glass-transition-like crossover without help from its surrounding solvent. Hence, in contrast to ordinary glasses the glass transition is not an intrinsic property of proteins. Instead, the dynamic crossover and other properties of proteins are driven by motions in the solvent, although the conformational changes in the protein occur on substantially longer time-scales than the solvent relaxation. In Fig. 9 it is seen that the relaxation rates of the protein processes are typically $10^3 - 10^5$ times slower than the α-relaxation in the solvent, proving the complex nature of global protein fluctuations and the associated energy landscape.

4 Conclusions

In this chapter we have investigated the dynamical behaviour of various systems containing confined water. Results from dielectric spectroscopy show that the confined water exhibits a dynamic crossover at a certain temperature (typically around 200 K) from a low temperature Arrhenius behaviour to a VFT dependence at higher temperatures. This observation together with other experimental facts, such as the absence of a calorimetric glass transition and a symmetric shape of the dielectric relaxation peak, suggest that the observed main relaxation process at low temperatures is due to a local β-relaxation, whereas a merged α–β-relaxation is responsible for the dynamics at higher temperatures. We also show that the apparent fragile-to-strong transition for confined supercooled water that has been observed in a number of QENS and ^{1}H NMR diffusion measurements is most likely due to a decoupling of water diffusion from the structural relaxation process.

Our dielectric measurements on myoglobin in water–glycerol mixtures of different water contents suggest that the most local protein motions are determined by the local β-relaxation of the surrounding water in the hydration shell, whereas more large-scale protein motions seem to be driven by large scale motions in the surrounding solvent. Thus, the present findings support the recently suggested idea that protein motions are solvent slaved [46, 64]. Moreover, we have shown that the relative amount of glycerol in the solvent determines the nature of both solvent and protein motions at low temperatures, which results in that at high glycerol contents both motions of the solvent and the protein seem to be of global character, while at high water concentrations the motions are of more local nature (although present at lower temperatures compared to the glycerol-rich solvents).

Acknowledgements We thank Prof. Frauenfelder for valuable discussions. This work was financially supported by the Swedish Research Council and the Swedish Foundation for Strategic Research.

References

1. F. Franks, *Water: A matrix of life,* 2nd ed., 2000, Cambridge: Royal Society of Chemistry.
2. Mathews, v. Holde, and Ahern, eds., *Biochemistry*, 3rd ed., 2000, Addison Wesley Longman: San Fransisco.
3. F. Franks, *Water: A comprehensive treatise*, ed. F. Franks, Vol. 1, 1972.
4. S. Zumdahl, ed., *Chemical principles*, 4th ed., 2002, Houghton Mifflin Company: Boston.
5. A. L. Lehninges, D. L. Nielsun, and M. M. Cox, eds., Principles of Biochemistry, 2nd ed., 1993, Worth Publishers: New York.
6. H. E. Stanley, *Unsolved mysteries of water in its liquid and glass states.* Mrs Bulletin, 1999. **24**(5): pp. 22–30.
7. M. C. Bellissent-Funel, *Water near hydrophilic surfaces.* Journal of Molecular Liquids, 2002. **96–7**: pp. 287–304.
8. P. Wernet, D. Nordlund, U. Bergmann, M. Cavalleri, M. Odelius, H. Ogasawara, L. A. Nslund, T. K. Hirsch, L. Ojamae, P. Glatzel, L. G. M. Pettersson, and A. Nilsson, *The structure of the first coordination shell in liquid water.* Science, 2004. **304**(5673): pp. 995–999.
9. S. R. Elliot, *Physics of amorphous materials*, 2nd ed., 1990, Longman Scientific and Technical, UK.
10. F. Kremer and A. Schönhals, eds. *Broadband Dielectric Spectroscopy*, 2003, Springer-Verlag.
11. G. P. Johari, *Intrinsic mobility of molecular glasses.* Journal of Chemical Physics, 1973. **58**(4): pp. 1766–1770.
12. H. Vogel, Phys. Z, 1921. **22**(645).
13. G. S. Fulcher, Journal of American Ceramic Society, 1925. **8**(789).
14. G. Tammann and G. Hesse, Zeitschrift fur Anorganische und Allgemeine Chemie, 1926. **156**(245).
15. C. A. Angell, *Formation of glasses from liquids and biopolymers.* Science, 1995. **267**(5206): pp. 1924–1935.
16. C. A. Angell, *Relaxation in liquids, polymers and plastic crystals - strong fragile patterns and problems.* Journal of Non-Crystalline Solids, 1991. **131**: pp. 13–31.
17. G. P. Johari, A. Hallbrucker, and E. Mayer, *The glass liquid transition of hyperquenched water.* Nature, 1987. **330**(6148): pp. 552–553.
18. A. Hallbrucker, E. Mayer, and G. P. Johari, *Glass-liquid transition and the enthalpy of devitrification of annealed vapor-deposited amorphous solid water – A comparison with hyperquenched glassy water.* Journal of Physical Chemistry, 1989. **93**(12): pp. 4986–4990.
19. C. A. Angell, *Liquid fragility and the glass transition in water and aqueous solutions.* Chemical Reviews, 2002. **102**(8): pp. 2627–2649.
20. V. Velikov, S. Borick, and C. A. Angell, *The glass transition of water, based on hyperquenching experiments.* Science, 2001. **294**(5550): pp. 2335–2338.
21. S. Cerveny, G. A. Schwartz, R. Bergman, and J. Swenson, *Glass transition and relaxation processes in supercooled water.* Physical Review Letters, 2004. **93**: pp. 245702.
22. M. Antognozzi, A. D. L. Humphris, and M. J. Miles, *Observation of molecular layering in a confined water film and study of the layers viscoelastic properties.* Applied Physics Letters, 2001. **78**(3): pp. 300–302.
23. P. Gallo, M. A. Ricci, and M. Rovere, *Layer analysis of the structure of water confined in vycor glass.* Journal of Chemical Physics, 2002. **116**(1): pp. 342–346.
24. M. O. Jensen, O. G. Mouritsen, and G. H. Peters, *The hydrophobic effect: Molecular dynamics simulations of water confined between extended hydrophobic and hydrophilic surfaces.* Journal of Chemical Physics, 2004. **120**(20): pp. 9729–9744.

25. S. Takahara, M. Nakano, S. Kittaka, Y. Kuroda, T. Mori, H. Hamano, and T. Yamaguchi, *Neutron scattering study on dynamics of water molecules in MCM- 41.* Journal of Physical Chemistry B, 1999. **103**(28): pp. 5814–5819.
26. U. Raviv, P. Laurat, and J. Klein, *Fluidity of water confined to subnanometre films.* Nature, 2001. **413**(6851): pp. 51–54.
27. M. A. Ricci, F. Bruni, P. Gallo, M. Rovere, and A. K. Soper, *Water in confined geometries: Experiments and simulations.* Journal of Physics-Condensed Matter, 2000. **12**(8A): pp. A345–A350.
28. M. Rovere and P. Gallo, *Effects of confinement on static and dynamical properties of water.* European Physical Journal E, 2003. **12**(1): pp. 77–81.
29. P. Pissis, J. Laudat, D. Daoukaki, and A. Kyritsis, *Dynamic properties of water in porous vycor glass studied by dielectric techniques.* Journal of Non-Crystalline Solids, 1994. **171**(2): pp. 201–207.
30. J. Swenson, R. Bergman, and S. Longeville, *A neutron spin-echo study of confined water.* Journal of Chemical Physics, 2001. **115**(24): pp. 11299–11305.
31. A. Faraone, L. Liu, C. Y. Mou, C. W. Yen, and S. H. Chen, *Fragile-to-strong liquid transition in deeply supercooled confined water.* Journal of Chemical Physics, 2004. **121**(22): pp. 10843–10846.
32. E. Mamontov, *Observation of fragile-to-strong liquid transition in surface water in CeO2.* Journal of Chemical Physics, 2005. **123**(17): pp. 171101.
33. K. Ito, C. T. Moynihan, and C. A. Angell, *Thermodynamic determination of fragility in liquids and a fragile-to-strong liquid transition in water.* Nature, 1999. **398**(6727): pp. 492–495.
34. J. Swenson, H. Jansson, and R. Bergman, *Relaxation processes in supercooled confined water and implications for protein dynamics.* Physical Review Letters, 2006. **96**(24): pp. 247802.
35. H. D. Middendorf, *Neutron studies of the dynamics of biological water.* Physica B, 1996. **226**(1–3): pp. 113–127.
36. N. Nandi and B. Bagchi, *Dielectric relaxation of biological water.* Journal of Physical Chemistry B, 1997. **101**(50): pp. 10954–10961.
37. V. P. Denisov and B. Halle, *Protein hydration dynamics in aqueous solution.* Faraday Discussions, 1996(103): pp. 227–244.
38. P. Gallo, M. Rovere, and E. Spohr, *Supercooled confined water and the mode coupling crossover temperature.* Physical Review Letters, 2000. **85**(20): pp. 4317–4320.
39. S. Dellerue and M. C. Bellissent-Funel, *Relaxational dynamics of water molecules at protein surface.* Chemical Physics, 2000. **258**(2–3): pp. 315–325.
40. M. Marchi, F. Sterpone, and M. Ceccarelli, *Water rotational relaxation and diffusion in hydrated lysozyme.* Journal of the American Chemical Society, 2002. **124**(23): pp. 6787–6791.
41. M. Tarek and D. J. Tobias, *Role of protein-water hydrogen bond dynamics in the protein dynamical transition.* Physical Review Letters, 2002. **88**(13): 138101.
42. A. L. Tournier, J. C. Xu, and J. C. Smith, *Translational hydration water dynamics drives the protein glass transition.* Biophysical Journal, 2003. **85**(3): pp. 1871–1875.
43. J. A. Rupley and G. Careri, *Protein hydration and function.* Advances in Protein Chemistry, 1991. **41**: pp. 37–172.
44. J. A. Rupley, P. H. Yang, and G. Tollin, *In water in polymers,* S. P. Rowland, ed., 1980, American Chemical Society: Washington D.C.
45. R. Pethig, *Protein-water interactions determined by dielectric methods.* Annual Review of Physical Chemistry, 1992. **43**: pp. 177–205.
46. P. W. Fenimore, H. Frauenfelder, B. H. McMahon, and R. D. Young, *Bulk-solvent and hydration-shell fluctuations, similar to α- and β-fluctuations in glasses, control protein motions and functions.* Proceedings of the National Academy of Sciences of the United States of America, 2004. **101**(40): pp. 14408–14413.
47. S. Cerveny, J. Mattsson, J. Swenson, and R. Bergman, *Relaxations of hydrogen-bonded liquids confined in two-dimensional vermiculite clay.* Journal of Physical Chemistry B, 2004. **108**(31): pp. 11596–11603.
48. S. N. Bhat, A. Sharma, and S. V. Bhat, *Vitrification and glass transition of water: Insights from spin probe ESR.* Physical Review Letters, 2005. **95**(23): 235702.

49. R. Bergman and J. Swenson, *Dynamics of supercooled water in confined geometry*. Nature, 2000. **403**(6767): pp. 283–286.
50. J. Swenson, H. Jansson, W. S. Howells, and S. Longeville, *Dynamics of water in a molecular sieve by quasielastic neutron scattering*. Journal of Chemical Physics, 2005. **122**(8): p. 084505.
51. H. Jansson and J. Swenson, *Dynamics of water in molecular sieves by dielectric spectroscopy*. European Physical Journal E, 2003. **12**: pp. S51–S54.
52. H. Jansson, R. Bergman, and J. Swenson, *Relation between solvent and protein dynamics as studied by dielectric spectroscopy*. Journal of Physical Chemistry B, 2005. **109**(50): pp. 24134–24141.
53. R. Bergman, J. Mattsson, C. Svanberg, G. A. Schwartz, and J. Swenson, *Confinement effects on the excess wing in the dielectric loss of glass-formers*. Europhysics Letters, 2003. **64**(5): pp. 675–681.
54. J. Swenson, G. A. Schwartz, R. Bergman, and W. S. Howells, *Dynamics of propylene glycol and its oligomers confined in clay*. European Physical Journal E, 2003. **12**(1): pp. 179–183.
55. J. Hedström, J. Swenson, R. Bergman, H. Jansson, and S. Kittaka, European Physical Journal, "Special Topics", 2007. **141**: pp. 53–56.
56. R. Bergman, J. Swenson, L. Börjesson, and P. Jacobsson, *Dielectric study of supercooled 2D water in a vermiculite clay*. Journal of Chemical Physics, 2000. **113**(1): pp. 357–363.
57. G. Sartor, A. Hallbrucker, K. Hofer, and E. Mayer, *Calorimetric glass liquid transition and crystallization behavior of a vitreous, but freezable, water fraction in hydrated methemoglobin*. Journal of Physical Chemistry, 1992. **96**(12): pp. 5133–5138.
58. J. Swenson, H. Jansson, J. Hedström, and R. Bergman, *Properties of hydration water and its role for protein dynamics*. Journal of Physics: Condensed matter, 2007. **19**(20): 205109.
59. L. Liu, S. H. Chen, A. Faraone, C. W. Yen, and C. Y. Mou, *Pressure dependence of fragile-to-strong transition and a possible second critical point in supercooled confined water*. Physical Review Letters, 2005. **95**(11): 117802.
60. S. H. Chen, L. Liu, E. Fratini, A. Faraone, and E. Mamontov, *Observation of fragile-to-strong dynamic crossover in protein hydration water*. Proceedings of the National Academy of Sciences of the United States of America, 2006. **103**(24): pp. 9012–9016.
61. F. Mallamace, M. Broccio, C. Corsaro, A. Faraone, U. Wanderlingh, L. Liu, C. Y. Mou, and S. H. Chen, *The fragile-to-strong dynamic crossover transition in confined water: Nuclear magnetic resonance results*. Journal of Chemical Physics, 2006. **124**(16): 161102.
62. L. Onsager and L. K. Runnels, *Diffusion and relaxation phenomena in ice*. Journal of Chemical Physics, 1969. **50**(3): pp. 1089–1103.
63. K. Goto, T. Hondoh, and A. Higashi, *Determination of diffusion coefficients of self-interstitials in ice with a new method of observing climb of dislocations by x-ray topography*. Japanese Journal of Applied Physics, **25**: pp. 351–357.
64. P. W. Fenimore, H. Frauenfelder, B. H. McMahon, and F. G. Parak, *Slaving: Solvent fluctuations dominate protein dynamics and functions*. Proceedings of the National Academy of Sciences of the United States of America, 2002. **99**(25): pp. 16047–16051.
65. D. Vitkup, D. Ringe, G. A. Petsko, and M. Karplus, *Solvent mobility and the protein 'glass' transition*. Nature Structural Biology, 2000. **7**(1): pp. 34–38.
66. W. Doster and M. Settles, *Protein-water displacement distributions*. Biochimica Et Biophysica Acta-Proteins and Proteomics, 2005. **1749**(2): pp. 173–186.
67. L. K. H. van Beek, *Progress in dielectrics*, ed. Heywood B. Birks, Vol. 7, 1967, London, pp. 69–114.
68. S. Bone, *Time-domain reflectometry studies of water binding and structural flexibility in chymotrypsin*. Biochimica Et Biophysica Acta, 1987. **916**(1): pp. 128–134.
69. Y. Shibata, A. Kurita, and T. Kushida, *Real-time observation of conformational fluctuations in Zn-substituted myoglobin by time-resolved transient hole-burning spectroscopy*. Biophysical Journal, 1998. **75**(1): pp. 521–527.
70. G. Careri, *Cooperative charge fluctuations by migrating protons in globular proteins*. 1998. Progress in Biophysics and Molecular Biology, **70**: pp. 223–249.

71. M. Ferrand, A. J. Dianoux, W. Petry, and G. Zaccai, *Thermal motions and function of bacteri-orhodopsin in purple membranes – Effects of temperature and hydration studied by neutron-scattering.* Proceedings of the National Academy of Sciences of the United States of America, 1993. **90**(20): pp. 9668–9672.

72. V. Kurkal, R. M. Daniel, J. L. Finney, M. Tehei, R. V. Dunn, and J. C. Smith, *Low frequency enzyme dynamics as a function of temperature and hydration: A neutron scattering study.* Chemical Physics, 2005. **317**(2–3): pp. 267–273.

73. J. H. Roh, V. N. Novikov, R. B. Gregory, J. E. Curtis, Z. Chowdhuri, and A. P. Sokolov, *Onsets of anharmonicity in protein dynamics.* Physical Review Letters, 2005. **95**(3): p. 038101.

74. H. Lichtenegger, W. Doster, T. Kleinert, A. Birk, B. Sepiol, and G. Vogl, *Heme-solvent coupling: A Mossbauer study of myoglobin in sucrose.* Biophysical Journal, 1999. **76**(1): pp. 414–422.

75. F. G. Parak, *Proteins in action: The physics of structural fluctuations and conformational changes.* Current Opinion in Structural Biology, 2003. **13**(5): pp. 552–557.

76. A. Huenges, K. Achterhold, and F. G. Parak, *Mössbauer spectroscopy in the energy and in the time domain, a crucial tool for the investigation of protein dynamics.* Hyperfine Interactions, 2002. **144**(1): pp. 209–222.

Interactions of Polarizable Media in Water and the Hydrophobic Interaction

F. Bresme and A. Wynveen

Abstract Water plays an essential role in determining the interactions between biological molecules in solution. A major component of these interactions is the hydrophobic force, the water-mediated interaction deemed to provide an impetus for the organization of living matter. Hydrophobic forces can be traced back to the dynamic and structural changes undergone by water at biological surfaces. Since biological surfaces are quite heterogeneous with regard to their interactions with water, rapid variations of the water properties at length scales on the order of molecular dimensions occur at these surfaces. This behavior is expected to be significant for proteins, where experiments show that the protein polarizability, as measured by the dielectric permittivity, can range from 4 for hydrophobic regions to 10 for regions of catalytic importance. These results emphasize the "patchy" character of these biomolecules regarding polarizability, an effect that may be relevant for understanding the conformational changes of proteins and other biomolecules in solution. In this chapter, we provide a critical discussion of the current state of hydrophobic interactions, with particular emphasis on biomolecules. Most of the theoretical approximations to date do not address specifically polarization effects. Polarization is, nonetheless, expected to play a major role in determining the interactions of biological water with uncharged surfaces. We discuss recent work that suggests that polarization can impart dramatic changes in the water behavior at hydrophobic surfaces and also can modify the strength of the hydrophobic interaction.

1 Introduction

The hydrophobic interaction is recognized as one of the main driving forces inducing the self-assembly of molecules in water solutions. Hence, it is responsible for the organization of surfactants and phospholipids into complex structures such as micelles

F. Bresme
Department of Chemistry, Imperial College London, SW7 2AZ, London, UK,
f.bresme@imperial.ac.uk

A. Wynveen
Department of Chemistry, Imperial College London, SW7 2AZ, London, UK,
a.wynveen@imperial.ac.uk

Bresme, F., Wynveen, A.: *Interactions of Polarizable Media in Water and the Hydrophobic Interaction*. Lect. Notes Phys. **752**, 43–62 (2008)
DOI 10.1007/978-3-540-78765-5_3   © Springer-Verlag Berlin Heidelberg 2008

and biological membranes [1]. One of the most common effects of hydrophobic forces is the demixing of oil-like compounds from aqueous solutions. The fundamental role played by this interaction in regulating conformational changes of proteins was discussed by Kauzmann in a very influential paper [2]. Tanford [3] also suggested that the self-assembly of complex biological structures is under thermodynamic control, and that hydrophobic forces can lead to biomolecular structures that may provide the impetus for the organization of living matter.

Experimental work has highlighted the connection between the solubility of hydrophobic units and biological processes such as protein folding. Privalov [4] performed extensive calorimetric studies of unfolding reactions of globular proteins. These experiments showed a large positive value of the heat capacity, as well as a characteristic temperature dependence of the enthalpy associated to protein folding. These experimental observations are attributed to the hydrophobic interaction. Experiments by Baldwin [5] on solutions of liquid hydrocarbons in water were used to explain the temperature dependence of the hydrophobic interaction on protein folding. Subsequent work by Spolar et al. [6] showed the existence of a correlation between the heat capacity and the water-accessible non-polar surface area. The analysis of these correlations in hydrocarbon compounds and globular proteins indicates a remarkable similarity between both systems. This result emphasizes the relevance of the hydrophobic interaction as a driving force in non-covalent protein processes, which are characterized by large changes in the heat capacity. The connection between hydrophobicity and protein folding has been revisited by Dill [7], who has noted that in addition to the hydrophobic interaction there are weaker forces (electrostatic repulsions, ion pairing, hydrogen bonding or van der Waals) that also affect the stability of proteins.

The hydrophobic effect can be traced back to the dynamic and structural changes undergone by water next to hydrophobic surfaces. The dynamic behavior of biological water next to protein surfaces has been investigated with femtosecond resolution [8]. These experiments show that water has a bimodal character, with some water molecules exhibiting bulk solvation times (≈ 1 ps) and other molecules displaying very slow dynamics (≈ 38 ps). The latter contribution can be associated with water that interacts strongly with the protein surface. On the other hand, nuclear magnetic resonance experiments of oxytocin in the unfolded state and the globular protein bovine pancreatic trypsin inhibitor have indicated the existence of two different types of hydration sites, with markedly different residence times, in the sub-nanosecond regime for water at the protein surface and between 10^{-2} and 10^{-8} s for proteins in the protein interior. Sum-frequency spectroscopy experiments [9] of water at hydrophobic surfaces (alkane interfaces) and computer simulations of hydrophobic nanoparticles [10, 11] have also furnished important structural information. The microscopic structure of water is consistent with a reduction of the number of hydrogen bonds per molecule. Also, the interactions with the hydrophobic phase results in a considerable orientation of water at the interface, which is characterized by the appearance of free OH bonds (dangling bonds) facing the hydrophobic surface. One might speculate that water behaves in a similar fashion at biological surfaces, although it is important to note that biomolecules are normally characterized by strong heterogeneities with regard to solvent–surface interactions. Recent simulation work [12] of the polypeptide mellitin in water indicate that water exhibits two types of structure: a clathrate structure, like those observed near convex surfaces, and a second structure compatible with the existence of dangling bonds that appears near flat surfaces. These two structures exhibit a substantial difference in the water–water interaction enthalpy.

The structural changes of the hydrogen-bond structure of water near a hydrophobic surface were already pointed out by Stillinger [13], who suggested that the three-dimensional hydrogen-bond network of water cannot be maintained next to a large hydrophobic solute (see Fig. 1). This has a direct consequence on the solvation free energy of hydrophobic solutes, demonstrating a functional dependence on the size of the solute. This dependence has been discussed by Chandler [14]. The solvation energy of large solutes (> 1 nm diameter) scales with the surface area, whereas that of smaller solutes scales with volume. With regard to the first case, a water depletion layer can be found next to hydrophobic solutes with small curvatures. The approach of such hydrophobic surfaces can result in capillary evaporation, which can generate strong attractive forces between the solutes.

There has been a great effort to explain the origin of the hydrophobic interaction. Nonetheless, there is still a need to understand the microscopic mechanism leading to this interaction and, in particular, the exact role played by water in determining the hydrophobic attraction. Several simulation studies of materials and biomolecules have tried to attain a molecular understanding of the mechanisms behind the hydrophobic effect [15, 16]. Complicating matters, however, recent simulations of biomolecules have offered contrasting results on the behavior of water in proteins and its relevance regulating protein folding [17, 18]. The main conclusion is that the role played by water in folding is extremely sensitive to the protein–water interactions. This result indicates that theoretical modeling of the hydrophobic effect may require considerable realistic properties of materials and biomolecules such as polarization effects. Polarization effects are generally ignored in most theoretical investigations of hydrophobicity, despite the fact that there is experimental and theoretical evidence that biomolecular systems can actually have significant permittivities [19, 20].

In this chapter we discuss the influence of intermolecular forces, in particular the polarization, on the hydrophobic interaction. The chapter is structured as follows: Firstly, we discuss theoretical approximations that have been proposed to explain hydrophobic attraction. These approximations rely to some extent on the notion of capillary evaporation, or drying, between hydrophobic surfaces. The existence of drying in these surfaces is nonetheless a matter of intense dispute in the literature.

Fig. 1 Sketch of the water structuring around small (*left*) and large solutes (*right*). For large solutes a depletion layer between the solute and the solvent appears

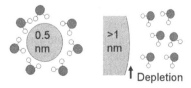

Thus, we provide a critical discussion on the feasibility of drying in materials and biological surfaces. We have made particular emphasis on the sensitivity of drying to biomolecule–water interactions, and we have intentionally biased part of our discussion toward the role played by solute polarization in modulating drying and the hydrophobic interaction. Polarization effects have systematically been neglected to certain extent in the investigations of hydrophobic substances. Nonetheless, as we discuss here, they might play a very important role in determining the interactions of uncharged surfaces in water. We finish our discussion with an outlook of future work in the investigation of the hydrophobic force.

2 The Origin of Strong Attraction Between Hydrophobic Surfaces

2.1 Drying as a Mechanism for the Hydrophobic Attraction

There is a large body of literature devoted to the physical origin of the hydrophobic interaction both in biophysics and in materials science. Experiments on hydrophobic surfaces have provided valuable results on the behavior of water under confinement. An understanding of the water behavior at hydrophobic materials can provide a basis to investigate the interactions of water with the complex surface landscapes appearing in biomolecular assemblies. Several experiments of water confined between hydrophobic surfaces [21, 22, 23] have reported the existence of strong long-range attractive forces between the surfaces. The hydrophobic surfaces are characterized by contact angles that are typically about 110°. More recent experiments [24] on hydrophobic surfaces have reported forces that are several times shorter-range than the ones measured in earlier experiments. Interestingly, the interactions between these hydrophobic surfaces conform to the classical capillarity and Lifshitz theories. The origin of the long-range attractive forces has been and still is a matter of debate. It was noted earlier that these long-range interactions cannot be described in terms of van der Waals forces alone, since the experimental forces are much stronger and their decay is exponential. In order to explain the decay length and strength of the forces, it was suggested that submicroscopic bridging cavities were responsible for the strong attraction. Thus, the process behind the attraction could be ascribed to cavitation or spontaneous drying between the surfaces at a specific inter-surface distance. This notion has been supported by recent experiments, which have reported the formation of nanobubbles at hydrophobic surfaces [25].

The idea of cavitation has been summarized in a general theory of solvation by Lum and co-workers [26]. This theoretical approach uses a mesoscopic square gradient theory to treat the water–vapor interface. It predicts large differences in the solvation of small and large solutes, and also, notably, it postulates large depletion regions around hydrophobic solutes and drying. Cavitation has also been reported in a number of computer simulation studies of water confined between hydrophobic surfaces [16, 27, 28, 29, 30, 31]. Also, it has been suggested that drying is responsible for the collapse of short polymer fragments [15] and that drying occurs in proteins [17, 18, 32]. Drying has also been considered to influence the permeation of water and ions through biological pores. Computer simulations of models of biological channels indicate that below a critical radius the water inside the pore evaporates, preventing ions from entering the pore [33]. This idea has been investigated in the context of density functional theory [34]. Although density functional theory provides a powerful theoretical framework to investigate phase transitions, it neglects the fluctuations associated to the small amounts of liquid moving into and out of the pore. These fluctuations are expected to play a significant role in the small confined spaces present in biological molecules. Also, it is important to note that the formation of a cavity is strongly dependent on the polarity of the pore [33], and under realistic conditions it could be restricted to pore sizes on the order of the water molecule radius. A systematic investigation of the effect of polarizability on hydrophobicity points in this direction [35].

Simple macroscopic thermodynamics offers an intuitive approach to understand the formation of cavities under confinement conditions. Most theoretical formulations have considered the grand potentials of the liquid, Ω_l, and the vapor, Ω_v, under confinement. For the case of two solutes the difference in the grand potential between one state where there is a cavity between the solutes and another state where there is liquid in between is given by,

$$\Delta\Omega = \Omega_v - \Omega_l = (P - P_v)V + 2\gamma_{lv}\cos\theta A_{sl} + 2A_{lv}\gamma_{lv} \tag{1}$$

where P is the pressure of the bulk liquid phase, P_v is the pressure inside the cavity, $\cos\theta = (\gamma_{sv} - \gamma_{sl})/\gamma_{lv}$ defines the contact angle of water with the hydrophobic surface, γ_{sv}, γ_{sl} and γ_{lv} are the surface tensions of the solute–vapor, solute–liquid and liquid–vapor interfaces, whose area is defined by A_{sl} and A_{lv}. Another contribution, the line tension, is normally ignored since it is a small force [36, 37]. Equation (1) can be used to find the critical distance defining the threshold for cavitation or water expulsion. This distance is geometry dependent [27, 28, 30, 35, 43]. For flat parallel surfaces of lateral size L the critical distance is given by

$$d_c \cong \frac{-2\gamma_{lv}\cos\theta}{(P - P_v) + b\gamma_{lv}/L} \tag{2}$$

where $b \approx 1$ is a geometric factor.

The force arising from the cavitation between two surfaces can be obtained by differentiation of the grand potential with respect to the inter solute separation D,

$$f = -\frac{\partial\Delta\Omega_{min}}{\partial D}\bigg|_T . \tag{3}$$

This yields a force due to cavitation on the order of $\approx 2\gamma_{lv}l$, where l is a characteristic length of the material or biomolecule in question. For water, $f/l \approx 15$ pN/Å, which is a rather strong force for nanoscopic-sized units.

The thermodynamic model discussed above has been tested in a number of systems using computer simulations. The model is accurate in predicting the qualitative behavior as well as the magnitude of the capillary force between hydrophobic objects [16, 27, 28, 30, 36]. With regard to biologically relevant problems, this simple theory has reproduced remarkably well the drying behavior of models of biological pores. Nonetheless, the accuracy of this model to predict drying in proteins (if present) has not been investigated in detail [32]. Unlike the well-defined geometrical models investigated in other computer simulations, a real protein has a very intricate shape, making it difficult to map the protein structure onto simple geometries.

2.2 Are Hydrophobic Surfaces Really Dry?

Despite all the theoretical and experimental investigations discussed above, the role played by drying in biological as well as material science problems is still a matter of dispute. This controversy has attracted considerable attention (see for instance [38]). Specifically, the formation of stable cavities bridging hydrophobic surfaces has not been established as a general feature of these systems [39]. Although cavitation has been reported in experiments using invasive methods such as atomic force microscopy [25, 40], it has been observed only at or close to contact with the surface force apparatus [39]. According to nucleation theory, the critical distance for cavity nucleation should be shorter than 3 nm [24], whereas some experiments have suggested the formation of cavities for surfaces lying tens of nanometers apart. Moreover, estimates of the free-energy barriers involved in cavity nucleation indicate that, in general, cavitation is an unlikely process for surface separations larger than typical molecular dimensions. Recent computations using well-defined slit pore models suggest that the free-energy barriers associated to cavitation in water can be as large as $20k_B T$ [41]. For such high-energy barriers cavitation might not be kinetically favored at ambient temperature, and metastable liquid states inside the confined region could appear instead. For low pressures (atmospheric pressure), theoretical models predict a strong dependence of the free-energy barrier on the contact angle, $\Delta G \propto 1/\cos\theta$ [42]. This relation has been tested with computer simulations of simplified models [43] showing good agreement with the theoretical expectations. The main message is that the cavitation process becomes more and more difficult as one approaches contact angles close to $90°$. Hence, only for contact angles of $180°$ (perfect hydrophobicity) would the process be spontaneous.

Very recent experiments of water next to hydrophobic surfaces are starting to provide a clearer image of the microscopic structure of water next to such surfaces. Neutron [44] and X-ray [45, 46] experiments have examined the interface between water and a hydrophobic layers (such as octadecylsilane). The results from these works support the existence of a layer of reduced density at the interface, i.e. they do not observe "nano-bubbles" or drying.

2.3 Sensitivity of Drying to Protein–Water Interactions

Understanding the forces underlying the folding of proteins as well as determining the folding temperature dependence is an important target of biophysics research. The exact role played by water continues to be a subject of intense investigation. In this sense the experiments and theoretical studies mentioned above can provide some important clues as to the behavior of water at hydrophobic surfaces in biological systems. It is important to note that the folding of soluble proteins is promoted by temperature ("cold denaturation") [47]. This suggests that the hydrophobic interactions become stronger with increasing temperature, a fact that is also observed in non-biological systems [14]. This observation is difficult to rationalize in terms of interfacial forces, since the surface tension decreases with increasing temperature. In fact an increase in the force with increasing temperature suggests that there is a strong entropic contribution to hydrophobic forces. Therefore, alternative explanations to the unusual temperature dependencies are sought.

In a paper, Pratt has reviewed the molecular theory of hydrophobicity [48]. In that review it is noted that the scaled particle theories have played a very important role in the understanding of hydrophobicity. These theories target the free energy associated with a hole forming in a liquid. The probability of generating a hole of a certain size is determined by the liquid structure, and it can be considered as a fingerprint of the liquid. At the molecular scale, water is stiffer than organic liquids (see [48]). This limits the probability to form holes much larger than the average hole size, which is of the order of one molecular diameter. In contrast, other solvents such as ionic liquids display a greater tendency to form holes of significant size [49]. It is interesting to note that theories designed to obtain information on the hole probability distribution in liquids predict some of the unusual dependencies of hydrophobic hydration with temperature [5].

Experiments also indicate that it might be difficult to generate dry states even after protein folding has occurred. Ernst et al. [19] investigated the presence of water in the solution structure of human interleukin-1β. They found evidence for the existence of positionally disordered water in a naturally occurring hydrophobic cavity. The actual behavior of water is nonetheless unclear, as observed in a considerable number of protein simulations, since drying inside a protein is very sensitive to the protein–water interactions [17, 18]. In one investigation [17] considering the BphC enzyme, it was observed that water remains in the structure with a density lower than the bulk density. When the electrostatic interactions were switched off drying was observed. One conclusion from these studies is that drying depends critically on the strength of the solute–solvent electrostatic interactions. A similar conclusion has been reached in computer simulations of polarizable and charged solutes immersed in water [30, 35, 50]. Alternative investigations of the mellitin tetramer have suggested the existence of a marked drying transition inside the tetramer structure, which consists of a nanoscale channel with a size of the order of two to three water molecule diameters. Nonetheless, it has been pointed out that such a drying transition is very sensitive to mutations in the protein residues [18]. Computer simulation studies of more simplified protein models based on Go's [51] potential indicate, however, that the folding can proceed in such a way that most of the structural transformation of the protein occurs before water is expelled [52]. The Go model reproduces many geometrical features of the transition state and the intermediates of rapidly folding proteins, and therefore it is expected that the results obtained with this model may be of relevance to a wide range of protein structures.

2.4 Water Behavior Near Polarizable Surfaces

The sensitivity of the drying process to the protein–water interaction suggests that drying, if present, is a very subtle process. It is important to note that drying would be favored in systems involving very weak solute–solvent interactions, i.e. for *very weakly* polarizable solutes and surfaces. There is experimental and theoretical evidence, however, that supports that the hydrophobic regions in proteins are actually polar to a certain extent. Experiments suggest permittivities of the order of 4 for hydrophobic spots [19] and $\epsilon \approx 10$ for regions of catalytic relevance. King et al. [20] have performed a detailed investigation of dielectric properties of proteins using molecular dynamics simulations. These authors investigated the electrostatics of trypsin and computed the dielectric constant at different regions in the proteins, finding a dependence of the permittivity on the protein site. This work emphasizes the usefulness of macroscopic concepts such as the dielectric constant to address the strength of electrostatic interactions in biomolecules. The existence of ionizable groups [53] in the interior of proteins lends further support to the existence of hydrophobic regions with medium permittivities.

Although most theoretical investigations of the hydrophobic interactions have completely ignored the polarizable character of hydrophobic solutes, some recent work has evaluated the impact of this interaction on biologically relevant systems [54]. The effect of polarization was considered in investigations into the passage of water through nanoscopic-sized channels, which could emulate protein channels. It was found that polarization can favor water permeation through the pore, and also intermittent permeation of water on nanosecond timescales as well as permeation enhancement due to the presence of ions inside the pore. The potential relevance of polarization in biological channels was already discussed by Parsegian more than 30 years ago [55]. Many theoretical approximations include the effect of dielectric discontinuities on surfaces (although many neglect, at first approximation, the solute and, notably, the solvent structures). For example, recent theories of DNA, using a Debye-Hückel form for the electrostatic interactions between DNA, found that short-range repulsive forces between DNA may be ascribed to the interaction between the helical charge on one molecule with its induced charge in the other molecule [56]. Such approaches have also been considered in investigations of electrostatic interactions in metalloproteins (cytochrome c_{551}, azurin and plastocyanin) [57]. This approach can provide important insights into the electrostatic interactions of these complex systems. Given the heterogeneous hydrophobic character of proteins [20] a more detailed microscopic representation of the dielectric inhomogeneities, as well as a molecular-level representation of the solute–water dielectric interface, is desired. Recent work [58, 59] has shown that water can exhibit an anomalous dielectric behavior near polar surfaces. This notion invalidates a simple representation of water as a dielectric media when considering highly confined systems.

Very recently we have developed a computational approach to investigate the interactions of polarizable solutes in explicit water. By using well-defined solute geometries it is possible to conduct a systematic investigation of the effect of polarizability on the water structure and on the solvation energies of the solute [30, 35]. This work suggests that polarization is indeed a very important variable in determining the physical behavior of water next to hydrophobic surfaces. Simulations of proteins with polarizable forcefields also point in this direction. Berne and co-workers have investigated, using molecular dynamics simulations, the bovine pancreatic trypsin inhibitor in water [60]. They find that the combined effect of protein and water polarizability results in an enhancement of water structure and dynamics near the protein surface. The main conclusion from both experimental and theoretical studies is that polarization cannot in general be ignored, and that a proper description of biological systems might require including its contribution, either explicitly or implicitly in an effective potential, if this is possible.

As outlined above, polarization can be an important element in the theoretical investigation of biological systems, and its contribution should not, in general, be neglected. Current questions of relevance in this area are: how might polarizability inhibit drying at a molecular scale?, what is the impact of polarizability on biological processes (e.g., protein folding)? and, generally, how does polarizability affect the hydrophobic interaction? These are of course very complex questions and answering them will require the concerted effort of experimentalists and theoreticians. In the next section we discuss recent developments in this direction.

3 Computer Simulation Treatment of Polarizable Hydrophobic Solutes

The investigation of the influence of polarizability on hydrophobic forces is a rather complex question. The development of new methodologies that enable the incorporation of solute polarization explicitly within computer simulations may help to address this question. There is a significant number of computer simulations that have considered polarization effects at the atomic level [60, 61, 62, 63, 64]. Many approximations assume a scalar polarizability and linear polarization. Thus the polarization effects have been included by computing the dipole induced by a given charge distribution on a particular atom [65]. Other approaches model the polarization site as a collection of charges whose magnitude can fluctuate in response to the environment [66]. An interesting implementation of this approach considers the charges as dynamic variables through a Lagrangian formalism [67].

The importance of polarization effects have been considered in the context of small hydrophobes, such as methane, in water. The first simulations of this system offered contrasting interpretations with regard to the influence of polarization on the methane–methane interactions [63, 64]. In a recent study it was shown that the interactions between these small solutes are rather insensitive to polarization. On the other hand computer simulations of extended polarizable surfaces [30, 35, 54] in explicit water indicate that the water structure is very sensitive to the surface polarizability. The observation of water restructuring at these surfaces indicates that the polarizability might also be relevant in determining the behavior of real hydrophobic surfaces. The modeling of the polarizability in extended surfaces has been done using an induced charge approach [30, 68]. This approach is briefly discussed in the following section.

3.1 Computer Simulations of Extended Polarizable Solutes

This methodology treats the polarizable solute as an isotropic linear dielectric whereas the solvent (water) is modeled explicitly. This means that the polarizability of the solute can be quantified in terms of a single permittivity, ϵ_{in}. In this way by varying ϵ_{in} one can study systematically the effect of solute polarizability. In order to compute the induced charges on the solute surface, a homogeneous set of grid charges that envelopes the solute is considered. The computation of the polarization reduces to a calculation of the magnitude of the grid charges for a given solvent charge distribution. This calculation requires the computation of the electrostatic potential through Poisson's equation,

$$\nabla^2 \Psi(r) = -\frac{q}{\epsilon_0} \delta(r - r') \tag{4}$$

which gives the potential at a point in space, r, due to a charge q located at r'. The total electrostatic potential everywhere, $\Psi(r)$ is given by,

$$\Psi(r) = \Psi_{ext}(r) + \Psi_{ind}(r) \tag{5}$$

where Ψ_{ext} and Ψ_{ind} are the electrostatic potentials due to the external charges and to the induced charges respectively. The external contribution can be computed using Coulomb's law. On the other hand the calculation of the induced charge contribution is a classical problem in electrostatics, provided the solute has a well-defined geometry. An explicit expression for the potential can be derived by considering that the normal component of the displacement field, D, and the tangential component of the electrostatic field, E, have to be continuous across the dielectric discontinuity associated with the water–solute interface, i.e.,

$$(D_{out} - D_{in}) \cdot n = 0 \tag{6}$$

$$(E_{in} - E_{out}) \times n = 0 \tag{7}$$

where 'in' and 'out' refer to regions inside and outside the polarizable solute. By imposing these conditions and assuming a sharp boundary between both dielectric media (solvent and solute) one can obtain Ψ_{ind}, and therefore, the total electrostatic potential, Ψ. The induced charge at the solute surface is then given by,

$$\sigma_{ind}(r_{surf}) = \boldsymbol{P} \cdot \boldsymbol{n} = -\epsilon_0 \chi \left(\nabla \Psi(r)_{r=r_{surf}} \right) \cdot \boldsymbol{n} \tag{8}$$

where χ is a function of the electric permittivities of the solute and the surrounding media (the solvent),

$$\chi = \frac{\epsilon_{in}}{\epsilon_{out}} - 1, \tag{9}$$

r_{surf} is a vector that represents the location of the induced charge on the solute surface, $\boldsymbol{n}$ is the unit radial vector normal to the solute surface, and $\boldsymbol{P}(r)$ is the polarization. For a solute defined by a cylindrical geometry one can derive an explicit expression for the induced charge (8) [30, 69],

$$\sigma_{ind}(R_c, \phi, z) = \sum_{m=-\infty}^{\infty} e^{im\phi} \int_0^{\infty} dk \frac{(\kappa-1)q}{2\pi^2 R_c} \frac{I'(kR_c)K_m(kd)}{I_m(kR_c)K'_m(kR_c) - \kappa I'_m(kR_c)K_m(kR_c)} \cos(kz) \tag{10}$$

which now defines the induced charge in terms of modified Bessel functions, $\kappa = 1 + \chi$, the position of the induced charges inside the cylinder, R_c, the distance, d, to an external point charge, q, and the cylindrical coordinates ϕ, z.

Recently, we have implemented an efficient molecular dynamics algorithm which relies on the computation of the induced charges using equation (10) [30]. For a solute immersed in an aqueous solution, the simulation proceeds by computing at each time step the induced charges associated with a given charge distribution of the solvent. The method provides a good account of the exact potential generated by the induced charges (see Fig. 2). The small differences with the exact potential are connected with the use of a grid to compute the induced charges. One advantage of using such a grid is that the electrostatic interactions can be computed using a standard Ewald or particle mesh Ewald methods [70].

  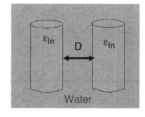

Fig. 2 (*Left*) Sketch of helix bundles in a lipid-binding protein (1EQ1), (*middle*) B-DNA chains, and (*right*) a model consisting of polarizable cylindrical solutes immersed in water. ϵ_{in} represents the permittivity of the hydrophobic solutes and D is the inter-solute distance

3.2 Water Structure Next to Extended Polarizable Solutes

The induced charge model described above has been used to simulate polarizable media immersed in water [30, 35, 54]. These simulations have considered a
rigid model for water, the Simple Point Charge Extended (SPC/E) model [71]. In
these studies the polarizability of water was not included. This begs the question
whether such an approximation is accurate. The SPC/E model very accurately predicts the bulk properties of water at ambient conditions, with thermodynamic and
structural properties similar to those obtained with more sophisticated polarizable
models [72]. Computer simulations of completely hydrophobic and metallic substrates offer similar conclusions [73, 74]. This contrasts somehow with the simulations of polarizable water next to a polarizable protein (bovine pancreatic trypsin
inhibitor) [60], in which qualitative differences in the behavior of water when both
the protein and water are polarizable are observed. Nonetheless, it is worth mentioning that the reported differences appear to be significant only near charged residues,
not hydrophobic residues. This fact makes it difficult to assess the impact of the
polarizability of a hydrophobic residue on the water properties. Nonetheless, using
simplified models of polarizable, but hydrophobic residues, this question may be addressed. Recently, we have investigated the behavior of water next to a cylindrical
polarizable solute. The cylindrical geometry is relevant to a number of biophysical
problems, such as the modeling of protein bundles or DNA–DNA interactions [56]
(Fig. 3). Similar models have also been used in theoretical approximations based on
the square gradient theory [26].

The simulations show that the water structure around the solute is very sensitive to the solute polarization. Figure 4 [30] shows a typical density profile of water

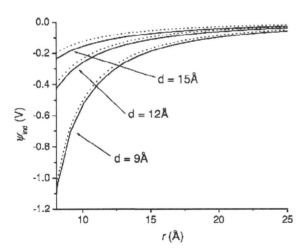

Fig. 3 Electrostatic potential due to an induced charge on a polarizable cylindrical solute ($\epsilon_{in} = 2$,
$R_c = 7$), arising from an external unit charge at different distance from the cylinder axis. *Full lines*
exact solution, *dotted lines* numerical approximation employed in molecular dynamics computer
simulations [30]

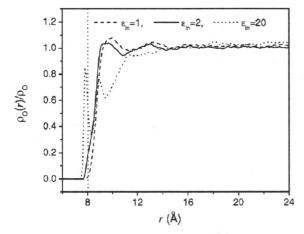

Fig. 4 Water (oxygen) density profile about a polarizable cylindrical solute. The graph illustrates the impact that the solute polarizability has on the water structure [30]

around a polarizable cylindrical solute. Non-polarizable solutes ($\epsilon_{in} = 1$) exhibit water depletion from the solute surface. Increasing the polarizability favors the adsorption of water on the solute. For relatively high permittivities ($\epsilon_{in} = 20$) the polarization induces strong layering in the water structure. These results suggest that regions of dissimilar permittivities, which are found in biomolecules such as proteins, will induce a very different water structuring. In addition to polarization, it is expected that the geometry and also the size (see for instance [75]) of the substrate will also play an important role in determining the structure and general behavior of water. Indeed, recent simulation work has shown that the protein topography can induce strong changes in the structure and free energy of hydration layers [12].

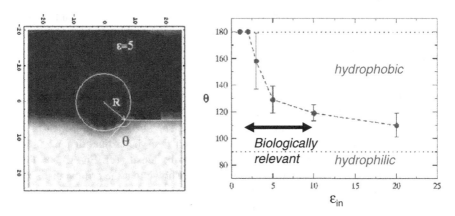

Fig. 5 (*Left*) Definition of the contact angle of a polarizable solute adsorbed at the water–air interface. *Light colors* indicate liquid density, *dark colors* vapor. (*Right*) Variation of the solute–water contact angle with solute polarizability (ϵ_{in})

An insight into the free-energy changes undergone by water near polarizable surfaces has been given recently [35]. Indirect information of surface energies (surface tensions) can be obtained by investigating the contact angle of water with polarizable solutes. Figure 5 shows an example of the dependence of the contact angle of a polarizable solute on the solute permittivity. For completely hydrophobic solutes that do not include attractive forces between the solvent and the solute, the contact angle is $180°$, i.e. there is complete drying. Interestingly, it was found that very weak polarizabilities, $\epsilon_{in} \approx 3$, can inhibit drying. Also, it was found that the wetting behavior of polarizable solutes varies significantly over a small interval of permittivities, $\epsilon_{in} = 1 \ldots 10$. This interval seems to be relevant to many biological systems, where normally hydrophobic regions are assigned permittivities in the range of $\epsilon = 1 \ldots 10$.

4 The Influence of Solute Polarizability on the Hydrophobic Interaction

Two questions of great biological significance are (i) what is the mechanism that drives hydrophobic attraction? and (ii) what is the order of magnitude of the forces arising from such interactions? As discussed above, although drying has been put forward as a possible explanation of the behavior, as observed in computer simulations of realistic proteins [17, 18, 32, 33], it is also true, as seen in these simulations, that drying is very sensitive to the solute–solvent interactions. We have shown recently that the hydrophobic interaction is likewise very sensitive to the solute polarizability, and that the microscopic mechanism associated with the strong attraction between weakly polarizable solutes cannot be explained in terms of capillary interactions alone.

Figure 6 shows an example of the variation of the hydrophobic forces between two cylindrical polarizable solutes with inter-solute separation. The main point to make is that the force is strongly dependent on the hydrophobic character of the solute, showing clearly a different behavior for purely hydrophobic substrates ($\epsilon = 1, \theta = 180°$) as compared with weakly polar solutes ($\epsilon > 5, \theta = 130°$). The hydrophobic force for purely hydrophobic substances is characterized by a discontinuous jump at a specific inter-solute separation. This behavior stems from the existence of capillary evaporation or drying (see Sect. 2.1). For solute separations smaller than the critical distance the inter-solute region nucleates a vapor cavity (see Fig. 6), resulting in the formation of liquid–vapor interfaces. The surface tension then provides a strong attractive force. For separations larger than the critical distance, the interstitial region is filled with liquid at the water bulk density (cf. Fig. 6) and the effective force between the solutes is zero. We have shown that the critical distance can be accurately estimated using the thermodynamic model discussed in Sect. 2.1. The applicability of the thermodynamic model to other polarizable solutes ($\epsilon \approx 5$) is nonetheless limited. In these cases the forces decrease continuously with solute separation, demonstrating an absence of capillary drying for these weakly polarizable objects. Also, the forces observed in this instance are of the same order as that of

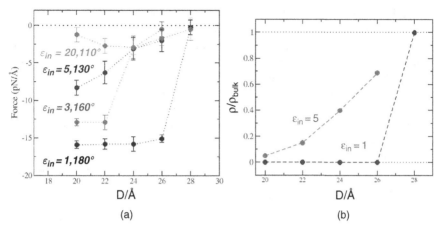

Fig. 6 (*Left*) Forces between polarizable solutes in water as a function of solute separation. The corresponding contact angles for the different permittivities is also indicated. (*Right*) Variation of the water density in the inter-solute gap with solute-solute separation

the capillary forces arising from drying, although the physical origin of these forces is completely different. In fact the attractive forces are connected to the existence of stable low-density states (see Fig. 6) in between the polarizable solutes. The forces arising from weakly polarizable objects can therefore be defined as *depletion forces*. We would like to note that the water behavior and the associated forces can be very sensitive to the pressure. This fact has to be taken into account in computer simulations of hydrophobic systems.

Figure 6 (*right*) illustrates the dependence of the density in the interstitial region with inter-solute separation. The density increases continuously with separation; this behavior is consistent with a continuous decrease in the force (cf. Fig. 6). The existence of a low density region at hydrophobic surfaces has been discussed in a number of theoretical and experimental works. Molecular dynamics simulations of the BphC enzyme have reported water densities that are $\approx 10\%$ the bulk water density. Similarly, low density states are found in water confined between structured hydrophobic substrates [29, 31]. Experiments also point toward the existence of low-density regions in the vicinity of hydrophobic surfaces. Neutron diffraction [44] and X-ray [45, 46] reflectivity experiments of water next to hydrophobic surfaces suggest the existence of a small hydrophobic gap of the order of the water molecule diameter. Although the reflectivity experiments do not correspond to confined systems, they provide evidence for the absence of bubble nucleation at hydrophobic surfaces. The simulation results discussed above are important for testing some hypotheses that have been put forward to explain the absence of cavities between hydrocarbon surfaces. The dependence of the activation energy for cavity nucleation on solute hydrophobicity, $\Delta G \propto 1/\cos(\theta)$, suggests the existence of a large kinetic energy barrier for cavitation, which would result in a liquid metastable state between hydrophobic surfaces [41, 43]. The results obtained recently for water confined between hydrophobic polarizable solutes offer another perspective (see Fig. 6, and [30, 35]). In this instance there is no evidence for metastable liquid states. The reproducibility of the low-density states, which are instead observed, can be tested within computer simulations [17, 35]. Figure 7a shows an example of such a test, where the low-density states are obtained irrespective of the initial conditions. Similar conclusions can be drawn from the results reported for the Bphc enzyme protein [17]. On the other hand, the simulations of water between polarizable cylinders (see Fig. 7a) also show that the water-density fluctuations in the interstitial region can be very significant. It has been shown that these fluctuations match a Gaussian distribution [35], indicating that the existence of intermittent liquid–vapor phases inside the confined region is not a likely explanation for the low-density phases. The importance of thermal fluctuations in the density of water confined between polarizable solutes can be inferred from the difference between the excess chemical potential, i.e., that attributed to interactions, of water in the interstitial region and of water in the bulk. The total chemical potential in the interstitial region, μ_{int}, is equal to the total chemical potential in the bulk, μ_{bulk}, since the interstitial region is an

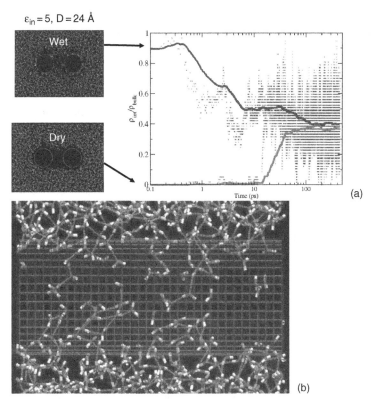

Fig. 7 (**a**) Evolution of water density in the inter-solute gap for two different starting configurations. (**b**) Snapshot of the water structure in the intersolute gap. Hydrogen bonds between water molecules have been highlighted as *dashed lines*

open region where the water molecules can enter and leave freely,

$$k_B T \ln \rho_{\text{int}} + \mu_{\text{int}}^{\text{ex}} = k_B T \ln \rho_{\text{bulk}} + \mu_{\text{bulk}}^{\text{ex}} \qquad (11)$$

For the low-density phases this results in an excess chemical potential difference of $\mu_{\text{int}}^{\text{ex}} - \mu_{\text{bulk}}^{\text{ex}} = -k_B T \ln[\rho_{\text{int}}/\rho_{\text{bulk}}] = +0.5 \cdots +2k_B T$, demonstrating the importance that thermal fluctuations play in the behavior of water in these confined systems.

Finally, the structure of the low-density phases described above seems to be compatible with the formation of fluctuating water aggregates, where the molecules form a low-dimensional network, characterized by a number of hydrogen bonds per molecule on the order of 2. In this sense hydrogen bonding represents a viable energetic alternative to stabilize these states. An example of such a low dimensional network is given in Figure 7b. It is worth mentioning that recent experiments on the structure and thermodynamics of the binding of solvent at internal sites in T4 lysozyme [76] suggest that water binding becomes more favorable in larger cavities, since the water molecules are free to form more hydrogen bonds between each other than in a less-confined space. The stability of water structures in hydrophobic cavities has also been illustrated in computer simulations of water in nanotubes. These simulations show spontaneous and continuous filling of hydrophobic carbon nanotubes [77]. In this case the water molecules form a one-dimensional ordered chain and develop densities inside the nanotube much higher than the bulk water density. This result might be of relevance to understanding water behavior in biological channels. Also, recent investigations of water confined in non-polar cavities made of fullerenes has shown that the fullerenes can be filled with highly structured water clusters [78]. Overall, all these studies indicate that water is very efficient in stabilizing states with densities different from the bulk value. This stabilization is very likely facilitated by the ability of water to form hydrogen bonds, which are energetically stronger than the typical dispersion forces operating in simple liquids.

5 Outlook

There is general consensus that hydrophobic interactions play a very important role in regulating biological processes such as protein folding and the self-assembly of biomolecules and materials. Nonetheless, there are many different interpretations concerning the microscopic mechanism regulating hydrophobic forces, and the exact details of how water regulates conformational changes is still a matter of debate. In the last few years, there have been important advances in our understanding of the molecular theory of hydrophobic effects. This has been possible thanks to developments of experimental techniques, advances in theoretical approaches as well as new computer simulation methodologies. In particular, computer simulations have provided unprecedented detailed information on the molecular structure of water confined within hydrophobic surfaces. Both simulations of biomolecules and materials have provided evidence for capillary evaporation of water under confinement. Nonetheless, they have also shown that drying is strongly dependent on the water–substrate interactions. Also, there is evidence that the topography of the substrate plays a relevant role in determining the local structure and energetics of water. Investigations of proteins and biological channels confirm this. Furthermore, detailed investigations of the effect of polarizability on nanosized substrates indicate that polarizability can inhibit drying. All these results do suggest that drying, if present, might appear only under rather exceptional circumstances in nanoscopic substrates.

There is growing evidence that, under confinement conditions, water can become stabilized in low-density states. These states do appear in the presence of hydrophobic substrates. Examples of such reduced density regions have been reported in computer simulations of proteins, hydrophobic pores, and extended polarizable surfaces. Likewise, neutron diffraction experiments and X-ray experiments of water at hydrophobic surfaces provide evidence that supports the existence of a depletion layer of molecular dimensions, i.e., there is no evidence for the formation of nanobubbles at the surface. Computer simulation investigations of polarizable hydrophobic objects offer a microscopic view consistent with a low- density phase that can undergo significant density fluctuations. Hydrogen bonding seems to play an important role in stabilizing these low-density states. More work is needed to understand the influence of solute geometry and solute–solvent interactions on the behavior of water in confined spaces. Also, further work is needed to investigate the relevance of these fluctuations in driving biomolecules through a particular kinetic path. It has been noted that water fluctuations can modify protein dynamics and even affect its function, providing another question for future study.

The sensitivity of hydrophobicity to the substrate–solvent interactions is reflected also in the forces between hydrophobic objects. It has been found very recently that substrate polarization has a very strong impact on the hydrophobic interaction. This modification of the forces is connected with the inhibition of drying and the stabilization of the low-density states mentioned above. As compared with the forces expected from capillary drying, the forces associated with low-density states are weaker but, surprisingly enough, not much weaker. In fact the typical forces measured in computer simulations, 25–50 pN, are of the same order of magnitude as the forces measured in experiments involving protein unfolding events [79, 80]. Further work is needed to understand the role played by water in tuning the forces measured in protein experiments. It is clear that a combination of powerful theoretical and experimental techniques will enable researchers to advance these complex questions in the forthcoming years.

Acknowledgements The authors would like to thank the EPSRC (FB) and The Royal Society (AW) for financial support. Computer resources on HPCx was provided via the UK's HPC Materials Chemistry Consortium (EPSRC grant EP/D504872). The authors would like to acknowledge the Barcelona Supercomputer Center (Spain) for providing resources on the Mare Nostrum Supercomputer, the Imperial College High Performance Computing Service, and the London eScience Centre at Imperial College London (UK) for providing computational resources.

References

1. J. Israelachvili: *Intermolecular and Surface Forces*, 2nd edn. (Academic, New York 1991)
2. W. Kauzmann: Adv. Protein Chem. **14**, 1 (1959)
3. C. Tanford: Science **200**, 4345 (1978)
4. P.L. Privalov: Adv. Protein Chem. **33**, 167 (1979)
5. R.L. Baldwin: Proc. Natl. Acad. Sci. U.S.A **83**, 8069 (1986)
6. R.S. Spolar, J.-H. Ha and M.T. Record: Proc. Natl. Acad. Sci. U.S.A **86**, 8382 (1989)
7. K.A. Dill: Biochemistry **29**, 7133 (1990)
8. S.K. Pal, J. Peon and A.H. Zewail: Proc. Natl. Acad. Sci. U.S.A **99**, 1763 (2002)
9. L.F. Scatena, M.G. Brown and G.L. Richmond: Science **292**, 908 (2001)
10. K.A. Tay and F. Bresme: J. Matter Chem. **16**, 1956 (2006)
11. K.A. Tay and F. Bresme: J. Am. Chem. Soc. **128**, 14166 (2006)
12. Y.K. Cheng and P.J. Rossky: Nature (London) **392**, 696 (1998)
13. F.H. Stillinger: J. Solution Chem. **2**, 141 (1973)
14. D. Chandler: Nature **437**, 640 (2005)
15. P.R. ten Wolde and D. Chandler: Proc. Natl. Acad. Sci. U.S.A **99**, 6539 (2002)
16. X. Huang, R. Zhou and B.J. Berne: J. Phys. Chem. B **109**, 3546 (2005)
17. R. Zhou, X. Huang, C.J. Margulis and B.J. Berne: Science **305** 1605 (2004)
18. P. Liu, X. Huang, R. Zhou and B.J. Berne: Nature (London) **437** 159 (2005)
19. J.A. Ernst, R.T. Clubb, H.X. Zhou, A.N. Gronenborn and G.M. Clore: Science **305** 1605 (2004)
20. G. King, F.S. Lee and A. Warshel: J. Chem. Phys. **95**, 4366 (1991)
21. R.M. Pashley et al.: Science **229**, 1088 (1985)
22. H.K. Christenson and P.M. Claeson: Science **239**, 390 (1988)
23. J.L. Parker, P.M. Claeson and P. Attard: J. Phys. Chem. **98**, 8468 (1994)
24. V. Yaminski and S. Ohnishi: Langmuir **19**, 1970 (2003)
25. J.W.G. Tyrell and P. Attard: Phys. Rev. Lett. **87**, 176104 (2001)
26. K. Lum, D. Chandler, and J.D. Weeks: J. Phys. Chem. B **103**, 4570 (1999)
27. X. Huang, C.J. Margulis and B.J. Berne: Proc. Natl. Acad. Sci. U.S.A **100**, 11953 (2003)
28. J. Dzubiella and J.P. Hansen: J. Chem. Phys. **121**, 5514 (2004)
29. N. Choudhury and B.M. Pettitt: J. Am. Chem. Soc. **127**, 3556 (2005)
30. A. Wynveen and F. Bresme: J. Chem. Phys. **124**, 104502 (2006)
31. N. Giovambattista, P.J. Rossky and P.G. Debendetti: Phys. Rev. E **73**, 041604 (2006)
32. Q. Huang, S. Ding, C.Y. Hua et al.: J. Chem. Phys. **121** 1969 (2004)
33. O. Beckstein and M.S.P. Sansom: Phys. Biol. **1**, 42 (2004)
34. R. Roth and K.M. Kroll: J. Phys. Condens. Matter **18**, 6517 (2006)
35. F. Bresme and A. Wynveen: J. Chem. Phys. **126**, 044501 (2007)
36. F. Bresme and N. Quirke: Phys. Rev. Lett. **80**, 3791 (1997)
37. F. Bresme and N. Quirke: J. Chem. Phys. **110**, 3536 (1999)
38. P. Ball: Nature **423**, 25 (2003)
39. H.K. Christenson and P.M. Claesson: Adv. Colloid Interf. Sci. **91**, 391 (2001)
40. R. Steitz, T. Gutberlet, T. Hauss et al.: Langmuir **19**, 2409 (2003)
41. K. Leung, A. Luzar and D. Bratko: Phys. Rev. Lett. **90**, 065502 (2003)
42. V.V. Yaminski, S. Ohnishi and B. Ninham: Long-range hydrophobic forces due to capillary bridging. In Nalwa, H.S., ed. *Handbook of Surfaces and Interfaces of Materials*, vol. 4, Solid thin films and layers. (Academic, New York 2001)
43. A. Luzar: J. Phys. Chem. **108**, 19859 (2001)
44. D.A. Doshi, E.B. Watkins, J.N. Israelachvili and J. Majewski: Proc. Natl. Acad. Sci. U.S.A **102**, 9458 (2005)
45. M. Mezger, H. Reichert, S. Schröder et al.: Proc. Natl Acad. Sci. U.S.A **103**, 18401 (2006)
46. A. Poynor, L. Hong, I.K. Robinson and S. Granick: Phys. Rev. Lett. **97**, 266101 (2006)
47. P.L. Privalov: Crit. Rev. Biochem. Mol. Biol. **25**, 281 (1990)
48. L.R. Pratt: Annu. Rev. Phys. Chem. **53**, 409 (2002)

49. F. Bresme and J. Alejandre: J. Chem. Phys. **118**, 4134 (2003)
50. J. Dzubiella and J.P.Hansen: J. Chem. Phys. **119**, 12049 (2003)
51. Y. Ueda, H. Taketomi and N. Go: Biopolymers **17** 1531 (1978)
52. M.S. Cheung, A.E. Garcia and J.N. Onuchic: Proc. Natl. Acad. Sci U.S.A **99** 685 (2002)
53. A. Warshel, S.T. Russell and A. Churg: Proc. Natl. Acad. Sci. U.S.A **81**, 4785 (1984)
54. R. Allen, S. Melchionna and J.P. Hansen: Phys. Rev. Lett. **89**, 175502 (2002)
55. A. Parsegian: Nature (London) **221**, 844 (1969)
56. A.A. Kornyshev and S. Leikin: J. Chem. Phys. **107**, 3656 (1997)
57. G. Iversen, Y.I. Kharkats and J. Ulstrup: Mol. Phys. **94**, 297 (1998)
58. J. Faraudo and F. Bresme: Phys. Rev. Lett. **94**, 077802 (2005)
59. J. Faraudo and F. Bresme: Phys. Rev. Lett. **92**, 236102 (2004)
60. B. Kim, T. Young, E. Harder, R.A. Friesner and B.J. Berne: J. Phys. Chem. B **109**, 16529 (2005)
61. I.M. Svishchev, P.G. Kusalik, J. Wang and R.J. Boyd: J. Chem. Phys. **105**, 4742 (1996)
62. G. Lamoureux and B. Roux: J. Chem. Phys. **119**, 3025 (2003)
63. M.H. New and B.J. Berne: J. Am. Chem. Soc. **117**, 7172 (1995)
64. D. Van Belle and S.J. Wodak: J. Am. Chem. Soc. **115**, 647 (1993)
65. L.X. Dang, J.E. Rice, J. Cadwell and P.A. Kollman: J. Am. Chem. Soc. **113**, 2481 (1991)
66. M. Sprik and M.L. Klein: J. Chem. Phys. **89**, 7556 (1988)
67. S.W. Rick, S.J. Stuart and B.J. Berne: J. Chem. Phys. **101**, 6141 (1994)
68. R. Allen, J.P. Hansen and S. Melchionna: Phys. Chem. Chem. Phys. **3**, 4177 (2001)
69. J.D. Jackson: *Clasical Electrodynamics*, 3rd edn. (Wiley, New York 1999)
70. M.P. Allen and D.J. Tildesley, *Computer Simulation of Liquids* (Clarendon Press, Oxford, 1989)
71. H.J.C. Berendsen, J.R. Grigera and T.P. Straatsma: J. Phys. Chem. **91**, 6269 (1987)
72. P.G. Kusalik and I.M. Svishchev: Science **265**, 1219 (1994)
73. A. Wallqvist: Chem. Phys. Lett. **165**, 437 (1990)
74. A. Kohlmeyer, W. Witschel and E. Spohr: Chem. Phys. **213**, 211 (1996)
75. N. Choudhury and B.M. Pettit: J. Am. Chem. Soc. **129** 4847 (2007)
76. J. Xu, W.A. Baase, M.L. Quillin, E.P. Baldwin and B.W. Matthews: Protein Sci. **10**, 1067 (2001)
77. G. Hummer, J.C. Rasaiah and J.P. Noworyta: Nature **414**, 188 (2001)
78. S. Vaitheeswaran, H. Yin, J. C. Rasaiah and G. Hummer: Proc. Natl. Acad. Sci. U.S.A **101**, 17002 (2004)
79. J.M. Fernandez and H. Li: Science **303**, 1674 (2004)
80. C. Cecconi, E.A. Shank, C. Bustamante and S. Marqusee: Science **309**, 2057 (2005)

Part II
Protein and Biological Solutions

Metastable Mesoscopic Phases in Concentrated Protein Solutions

P.G. Vekilov, W. Pan, O. Gliko, P. Katsonis and O. Galkin

Abstract It is sometimes claimed that the cytosol around the organelles, tubules, and other cellular structures represents a liquid phase. On the other hand, almost all protein molecules in the cytosol participate in complexes with other proteins, nucleic acids, small molecules, etc. The two pictures of a homogeneous liquid and a granular multiscale mixture appear incompatible. Thus, an important question in physical biology is whether the protein complexes represent a property of the protein solutions, or are the result of complex specific interaction involving multiple biological molecules.

We apply light scattering, atomic force microscopy, and other techniques to demonstrate that even solutions of a single protein of moderate concentration do not comply with Gibbs's definition of phase. In such solutions clusters of sizes from several tens to several hundred nanometers exist and have limited lifetimes. These clusters have a higher free energy than the protein solution, and their lifetime is determined by a barrier for their decay. The clusters affect the viscous and viscoelastic behavior of the solution and are an essential part of potential condensation and aggregation pathways. Since the clusters are observed in solutions of single proteins, they indicate that the proteins have an intrinsic propensity to form mesoscopic structures, which likely is utilized in the formation of the protein complexes in the cytosol.

P.G. Vekilov
Department of Chemical and Biomolecular Engineering, and Department of Chemistry, University of Houston, Houston, TX 77204, USA, pgvekilo@mail.uh.edu

W. Pan
Department of Chemical and Biomolecular Engineering, University of Houston, Houston, TX 77204, USA, wpan@mail.uh.edu

O. Gliko
Department of Chemical and Biomolecular Engineering, University of Houston, Houston, TX 77204, USA, ogliko@mail.uh.edu

P. Katsonis
Department of Chemical and Biomolecular Engineering, University of Houston, Houston, TX 77204, USA, pkatsonis@mail.uh.edu

O. Galkin
Department of Chemical and Biomolecular Engineering, University of Houston, Houston, TX 77204, USA, ogalkin@mail.uh.edu

Vekilov, P.G. et al.: *Metastable Mesoscopic Phases in Concentrated Protein Solutions.* Lect. Notes Phys. **752**, 65–95 (2008)
DOI 10.1007/978-3-540-78765-5_4

Cluster theories developed for colloid systems appear inapplicable to proteins due to the high level of implied Coulombic repulsion. A Monte Carlo model with protein-like potentials reproduces the metastable clusters of dense liquid with limited lifetimes and variable sizes, and suggests that the clusters' sizes are determined by the kinetics of growth and decay, and not by thermodynamics. A microscopic theory, which should account for stabilizing and destabilizing factors involving protein molecules and solvent inside the clusters, is still to be developed.

1 Macroscopic and Mesoscopic Phases

The cytoplasm of living cells is a highly heterogeneous environment, containing organelles, tubules, filaments as well as the "free" cytosol [1]. This free cytosol is an aqueous solution of proteins, RNA, and smaller organic and inorganic molecules. The components of the cytosol occupy from 30 to 50% of its volume [2]. Besides being crowded, the "free cytosol" is neither homogeneous nor at a steady state: The sizes of the molecules in it vary from 2 Å for water to several tens of nanometers for macromolecular complexes. Many of these complexes have finite lifetimes and characteristic times of their formation and decay occupy a wide spectrum. The most fundamental processes of life—biochemical transformations, binding, protein synthesis and folding, aggregation, etc.—take place in this crowded and structurally and dynamically complex environment.

This complexity of the cytosol is reflected in the ongoing discussion about its phase state: while it is generally accepted that the cytosol is a liquid, there are claims that it is more akin to a gel [3]. Another controversy surrounds the issues of whether the cytosol is a single phase, or represents multiple phases in equilibrium [4]. If one attempts to understand this controversy from first principles, one should start from Gibbs's definition of phase. A phase was defined as a homogeneous part of a heterogeneous system [5, 6]. The homogeneity was assumed to only apply to macroscopic lengthscales, while at certain mesoscales, fluctuations of properties (composition, density, structure, etc.) occur [5, 6, 7]. Importantly, the deviations from the mean values of the properties in a fluctuation were assumed to be infinitesimal. Clearly, the cytosol does not fit the classical definition of phase.

With that in mind, a question arises if the inhomogeneities, structures, and dynamics in the cytosol are intrinsic properties of the macromolecular solutions, or these features only appear if multiple proteins and nucleic acids interact via complex biological mechanisms. In the former case, structures would be present even in solutions of single proteins. One can start to address this issue from recent results on colloid solutions, resembling solutions of proteins. Theory and simulations predict that the combination of Coulomb repulsion at intermediate separations and attraction via, e.g., van-der Waals forces, at short separations, could lead to the formation of clusters [8, 9, 10]. The lifetime of these clusters was found to be long and they were assumed to be in equilibrium with the initial solution, i.e., stable. However, their composition was found to differ from any microscopic phase, present in the phase diagram in the solution. Furthermore, over long periods, the clusters were *not* found to grow to macroscopic regions of another phase. Thus, the system with the clusters *is not* the heterogeneous system with phases represented in the phase diagram [10, 11].

Particularly intriguing are clusters in protein solutions. If such clusters were found, this would indicate that the proteins have an intrinsic propensity for aggregation and association, which is likely the basis for the dynamic structures existing in the cytosol. Another reason why protein clusters are of interest is because they may be involved in phase transitions in the solution. Such phase transitions—aggregation [12, 13, 14], crystallization [15], polymerization [16], fibrilization [13, 17], etc.—underlie several pathological conditions and are parts of laboratory and technological procedures [18].

An important example of the protein phase transitions with pathophysiological consequences is sickle cell anemia [19, 20]. This is a genetic disease caused by a single mutation from glutamate to valine at the sixth site of the hemoglobin β chain [21], resulting in abnormal hemoglobin, HbS [22]. Despite the local character of the mutation, the clinical manifestations of the disease are very diverse and range from life-threatening to asymptotic conditions [19, 20]. Although stem-cell transplantation and gene therapy are promising the hope of a cure [23], in practice these treatments may be costly and not readily available in developing countries, where the incidence of the disease is highest. Thus, a search for other means of therapeutic intervention is still ongoing [20].

One of the basic events in sickle cell anemia is the polymerization of HbS inside the red blood cells, which occurs upon deoxygenation of hemoglobin [24]. HbS polymerization is a first-order phase transition in solution [25]. Significant efforts invested in the investigation of the polymerization mechanisms have brought a wealth of information and a feeling of understanding of the physico-chemical fundamentals of the process [16], yet no effective antisickling drug has been found [26]. The current treatment strategies do not include attempts to directly inhibit hemoglobin sickling [27]. A major obstacle for the antisickling approach has been the high concentration of hemoglobin inside the erythrocytes, which is incompatible with acceptable concentrations of a drug targeting most HbS molecules [28]. Thus, further studies of the polymerization mechanisms are needed.

The process of HbS polymerization has been divided into homogeneous nucleation of polymer fibers, growth of the fibers, and heterogeneous nucleation of new fibers on top of existing ones, leading to spherulites and eventually to a thick gel [29, 30, 31]. The homogeneous nucleation has an exponential dependence on HbS concentration and is susceptible to control via the solution composition. Recently, there has been an extensive discussion of a nucleation mechanism for ordered solids, whereby dense liquid phases serve as precursors and enhancers of nucleation [32, 33, 34, 35, 36, 37, 38, 39, 40]. In view of the lack of macroscopic dense liquid in near physiological HbS solutions, particularly important is the finding that such dense liquids carry out their nucleation function even if they are metastable with respect to the generating solution and only exist as mesoscopic clusters [36, 39, 40].

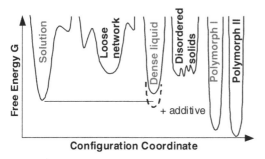

Fig. 1 Free energy G landscape for different phases possible in protein solutions. The abscissa is a generalized configuration coordinate, which can be thought of as protein concentration + degree of ordering of the protein molecules. For many protein solutions in certain ranges of concentration, temperature, and additive type and concentrations, the dense liquid phase is metastable with respect to both the protein solution at *left* and the ordered solids at *right*. Often, increasing the concentration of an additive lowers the free energy ΔG of the dense liquid to a level below that of the solution. Because of the slow formation of the ordered solids, the dense liquid is kinetically stabilized and can form macroscopic regions

In these lecture notes, we discuss dense liquid clusters in protein solutions. To understand the participation of the protein clusters in the formation of other phases, one should consider a free energy diagram, which includes clusters. In the free energy landscape [41] of a system with complex interactions, illustrated in Fig. 1, the homogeneous solution and an ordered solid phase (e.g., crystal) can be viewed as occupying deep valleys at the ends of a configurational coordinate, a one-dimensional projection of the multi-dimensional order parameter The configurational space between the minima is taken not by a single smooth ridge, but may possess several shallower valleys, separated by low ridges. These valleys represent metastable and disordered, including "liquid," phases. The heights of the ridges between them determine the lifetimes of the metastable phases. Within this viewpoint, the clusters appear as mesoscopic pieces of the intermediate metastable phases that never grow to macroscopic dimensions due to their limited lifetimes and are distinct from the phases occupying the deep minima.

The experimental tests for clusters below were carried out with two proteins: sickle cell hemoglobin and lumazine synthase, depicted in Fig. 2. In HbS solutions, pathophysiology-related polymerization occurs at concentrations above 170 mg ml^{-1} and this is why we study the clusters at concentrations in the range of 60–130 mg ml^{-1}. In solutions of lumazine synthase, crystallization occurs at a low 2.5 mg ml^{-1} [42]. Hence clusters were studied at concentrations between 1 and 9 mg ml^{-1}. To probe the effect on clustering of the mutation in HbS which leads to its pathological behavior, we also monitored clustering in solutions of oxy-HbS and oxy- normal adult hemoglobin, HbA. Oxy-HbS differs from deoxy-HbS by a conformation change from T- to R-(in oxy-HbS) state, which facilitates the binding of four O_2 molecules [43, p. 212]. Oxy-HbA differs from oxy-HbS by having a glutamate instead of valine at the two $\beta 6$ sites. The only condensed phases known for oxy-HbS and oxy-HbA are crystals that form over many days. Experimental details

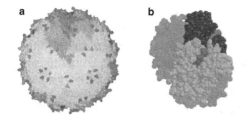

Fig. 2 Models of the protein molecules. (**a**). Lumazine synthase from *Bacillus subtilis*, an icosahedral hollow capsid of 60 identical subunits [45, 46], involved in the biosynthetic pathway of riboflavin [47, 48]. It consists of 60 identical subunits and has icosahedral symmetry with diameter $\sim$15.6 nm and molecular mass $Mw \sim 10^6$ g mol^{-1}. Note the roughness of the molecular surface. (**b**). Human hemoglobin, oxygen transporter [49]. The molecule consists of four subchains, two α and two β. Long molecular dimension is $\sim$5.5 nm, molecular mass Mw = 64,000 g mol^{-1}

about the studied proteins are provided in [42, 44]. Before we discuss the results with the two proteins, we introduce the experimental methods that can be used to study and understand the clustering behavior in protein solutions.

2 Methods of Detection and Monitoring of Metastable Clusters

2.1 Dynamic Light Scattering Monitoring of Clusters Sizes and Numbers

Dynamic light scattering (DLS) data are collected at angles θ ranging from 30 to 150°. The typical cuvettes for static and dynamic light scattering are cylindrical with a 10 mm diameter. These cuvettes require at least 400 µl of solution, which may not be available for a precious protein. One can fabricate small-volume cuvettes from a silica tube with an inside diameter of 1.5 mm to reduce the solution volume to $\sim$40 µl. The results obtained with the small cuvettes should be validated by a comparison to data acquired using a standard cuvette with an easily available protein.

To determine the lifetime of the clusters and monitor their long-term evolution, DLS data are collected every 30 s over periods of up to 3 h, starting within several seconds of solution preparation.

DLS yields an intensity correlation function $g_2(\tau)$, which characterizes the rate of intensity variations $I(t)$ over a total acquisition time Δt. g_2 is defined from the intensity at two times, t and $(t - \tau)$ as [50]

$$g_2(\tau) = \langle I(t)I(t-\tau)\rangle_{\Delta t} / \langle I\rangle^2_{\Delta t} \qquad (1)$$

where $\langle\rangle_{\Delta t}$ signifies averaging over time Δt and $\langle I\rangle_{\Delta t}$ is the average intensity for the entire period of data acquisition. Since light is scattered by the fluctuations of concentration, the correlation function mostly characterizes the rate of diffusion of the scatterers during the decay of the fluctuations [51]. This diffusion rate is, in turn, determined by scatterers' sizes and the viscosity of the medium, in which they are suspended [52]. In some cases, the fluctuations decay not because of the diffusion of scatterers, but because of decay of scatterers with limited lifetime. This case is discussed in relation to the clusters in hemoglobin solution below.

The normalized correlation function $g_2(\tau)$, illustrated in Fig. 3, can be represented as the square of the sum of exponential members representing the scatterers with different diffusion rates Γ_i, modified by a noise function ε. Our dynamic light scattering experiments are aimed at identifying one or two scatterers: single molecules and, in some cases, larger clusters. Hence [53]

$$g_2(\tau) - 1 = \left(A_1 e^{-\tau/\tau_1} + A_2 e^{-\tau/\tau_2}\right)^2 + \varepsilon(0, \sigma(\tau)) \tag{2}$$

where $\tau_1 = 1/\Gamma_1$ and $\tau_2 = 1/\Gamma_2$ are the characteristic times of the diffusion of scatterers (Γ_1 and Γ_2 are the respective diffusion rates), whose contribution to the scattered light has amplitudes A_1 and A_2. $\varepsilon(0, \sigma(\tau))$ is the noise function, which is expected to have a Gaussian distribution width σ, centered at 0.

The characteristic times τ_1 and τ_2 are more readily determined from the distribution function of the decay rates $G(\tau)$, illustrated in the same Fig. 3. This $G(\tau)$ can be expressed as

$$g_2(t) - 1 = \left(\int G(\tau) \exp(-t/\tau) d\tau\right)^2 \tag{3}$$

and is calculated by numerically inverting the Laplace transform with $(g_2 - 1)^{1/2}$ [54, 55].

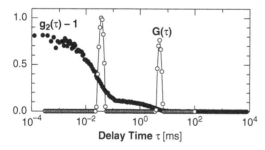

Fig. 3 Light scattering characterization of solutions of sickle-cell hemoglobin in deoxy-state (deoxy-HbS). Examples of correlation function $g_2(\tau)$ (*solid circles*) and the corresponding delay time distribution function $G(\tau)$ (*open circles*) of a deoxy-HbS solution with $C_{Hb} = 67$ mg ml^{-1} at $T = 25°$C. From [44]

To calculate the equivalent hydrodynamic radii from the relaxation times, the Stokes–Einstein relation is modified with $\Gamma_i = \tau_i^{-1} = D_i q^2$

$$R_i = \frac{k_B T\, q^2}{6\pi\eta_i}\,\tau_i \tag{4}$$

where $i = 1, 2$ for single molecules or clusters, respectively, k_B is Boltzmann constant, T is absolute temperature, η_i is the viscosity to which a diffusing object i is exposed, D_i is its diffusion coefficient, and the scattering vector $\mathbf{q} = 4\pi n/\lambda \sin(\theta/2)$ (n is the refractive index of the respective solvent, λ is the wavelength of the used laser).

Separate viscosity η_i should be used for the single protein molecules and for the clusters. For the single molecules, one takes the viscosity of the solvent in which the protein is dissolved, these are typically available in reference books [56]. One can correct for the temperature of the determination by assuming temperature dependence similar to water. Because the sizes of clusters are significantly greater than the size of the protein molecules, in calculations for the clusters we use the viscosity of the protein solution. For some proteins, such as hemoglobin, the viscosity is available even for high concentrations [57], for others it can be determined as discussed below.

To estimate the concentration of clusters n_2 and the fraction φ_2 of the solution volume occupied by them, one uses that the ratio of the amplitudes A_1 and A_2 in (2). As shown in [44], the final expressions for n_2 and φ_2 are

$$n_2 = \frac{A_2}{A_1}\frac{1}{P(\mathbf{q}R_2)\, f(C_1)}\frac{(\partial n/\partial C_2)_{T\cdot\mu}}{(\partial n/\partial C_1)_{T\cdot\mu}}\left(\frac{\rho_1}{\rho_2}\right)^2\left(\frac{R_1}{R_2}\right)^6 n_1 \tag{5}$$

$$\varphi_2 = \frac{A_2}{A_1}\frac{1}{P(\mathbf{q}R_2)\, f(C_1)}\frac{(\partial n/\partial C_2)_{T\cdot\mu}}{(\partial n/\partial C_1)_{T\cdot\mu}}\left(\frac{\rho_1}{\rho_2}\right)^2\left(\frac{R_1}{R_2}\right)^3 \varphi_1$$

In (5), the shape factor, assuming spherical shape of the clusters, is

$$P(\mathbf{q}R_2) = \left[\frac{3}{(\mathbf{q}R_2)^3}\left(\sin(\mathbf{q}R_2) - \mathbf{q}R_2\cos(\mathbf{q}R_2)\right)\right]^2, \tag{6}$$

$f(C_1)$ is a virial type expression accounting for intermolecular interaction between proteins molecules (the interactions between clusters are neglected because of low concentration). $(\partial n/\partial C_i)_{T,\mu}$ is the refractive index n increment, ρ_1 and ρ_2 are the protein densities in the single molecules and in the clusters.

The dynamic light scattering determinations of cluster sizes and concentration have errors due to two reasons: inaccuracies of the numerical inversion of the Laplace transform [58], (3), and noise in the system. For a relatively reliable characterization of the clusters, the use of a "multiple-tau" correlation technique is important. In this technique, the correlation function is simultaneously calculated with a set of several different lag times, τ. In contrast to systems with constant sampling times, this technique produces noise amplitude, which diminishes with increasing τ's. This reduces the error of measurements at long lag times, where the signal from the clusters is located.

An evaluation of the errors in dynamic lights scattering data, in which the second peak corresponds to clusters freely diffusing in the solution, has yielded the following conclusions: The second peak will be undetectable if it is below the noise level, typically around a few tenths of one percent. If the second peak has an amplitude higher than this threshold, it is readily detected, but the scatter in its position is high. As the amplitude of the second peak grows, the error in the determination of its position decreases. The position of the second peak can be determined with an error of $< 20\%$ if $A_2/A_1 \approx 0.05$ and the error drops to $\sim 10\%$ if $A_2/A_1 \approx 0.1$. The noise level can be lowered by increasing the total time of acquisition of the intensity variation trace, and by increasing the concentration of scattering molecules [44].

2.2 Static Light Scattering Characterization of the Intermolecular Interactions

Static light scattering (SLS) allows determination of the molecular mass of a protein M_w and the second osmotic virial coefficient B_2. The coefficient B_2 is a part of the factor f in (5) and is necessary in for the determination of cluster concentrations. More importantly, B_2 is used to characterize the interactions between protein molecules in solution. B_2 and M_w are extracted from the scattered intensity using experimentally determined Debye plots [59, 60].

$$\frac{KC}{R_\theta} M_w = 1 + 2B_2 M_w C \tag{7}$$

where $R_\theta = I/I_0$ is the Rayleigh ratio of the scattered to the incident light intensity, K is a system constant defined as [51] $K = N_A^{-1} \left(2\pi n_0/\lambda^2\right)^2 (dn/dC)^2$, where n_0 = 1.331 is the refractive index of the solvent at the wavelength of the laser beam λ [61], and N_A is Avogadro's number. Since often the magnitudes of B_2 are small, up to 10 or even 15 R_θ, C pairs should be used to minimize the error of the determination.

The second osmotic virial coefficient B_2 is defined as an integral over the whole space around a molecule of a function of the intermolecular interaction potential $U(r)$.

$$B_2 = \frac{2\pi N_A}{M_w^2} \int\limits_0^\infty \left\{1 - \exp\left[-U(r)/k_B T\right]\right\} r^2 dr \tag{8}$$

Clearly, a single value of B_2 cannot yield any details about the intermolecular interaction potential $U(r)$. However, if B_2 is known as a function of, e.g., the concentration of a crucial additive for the existence of clusters, the nature of the intermolecular forces contributing to B_2 can be qualitatively analyzed.

2.3 Determination of Solution Viscosity at Low-Shear Rates

Data on the viscosity of the solution in which clusters form are needed for the determination of the clusters radii. For these determinations, it is assumed that the viscosity is determined by the single molecules in the solution and is only weakly affected by the presence of the clusters. This may not always be true, especially at higher protein concentration, at which the clusters occupy more than 0.001 of the total volume and cluster interactions may become significant. Viscosity also helps to discriminate between two sources of slow processes reflected in the correlation function: the diffusion of the clusters and the diffusion of single molecules inside high-viscosity structures, such as loose networks and others. For the latter objective, viscosities at high and low shear are compared. In many cases, the high-shear viscosities are available.

Typical flow-through or dropping ball viscometers require large amounts of solution and are often inapplicable to tests with poorly available proteins. These viscometers operate at high-shear rates, where potential weakly bound structures are likely destroyed. An alternative method was developed based on dynamic slight scattering from polystyrene beads suspended in the protein solution [62].

Non-interacting spherical particles with diameter $2R_o = 390$ nm are mixed with the protein solution, and the correlation function $g_2(\tau)$ from this suspension is recorded. From $g_2(\tau)$ and the respective $G(\tau)$, the decay rate Γ_p of the particles is determined as discussed above. By using the Einstein–Stokes equation, the viscosity η of the solution is obtained from Γ_p as

$$\eta = \frac{k_{\mathrm{B}}T}{6\pi R_o} \frac{q^2}{\Gamma_p}. \tag{9}$$

A diffusing particle of 0.39 μm diameter covers its own size for ∼0.1 s. The thickness δ of the viscous boundary layer around it, at the top of which the solution velocity u vanishes, is of the order of 10 μm. The shear $s = \Delta u/\delta$ induced by the Brownian motion of such a particle is <1 s^{-1} [44].

2.4 Atomic Force Microscopy

The typical atomic force microscopy (AFM) instruments and methods of imaging in protein have been discussed in numerous papers [63, 64, 65]. The limitations of AFM for cluster imaging are that AFM images are collected over periods of several tens of seconds to several minutes. This is often longer than the clusters' lifetime. Thus, AFM has only been used to verify the existence of clusters in a solution using the fact that upon contact with a crystal surface, the clusters are stabilized and eventually evolve their structure into a crystal [42].

3 Intermolecular Interactions in Solutions of Lumazine Synthase and Hemoglobin

An understanding of protein cluster formation starts with insight into the protein intermolecular interactions. The ionic strength in the cytosol is $\sim$0.1–0.15 M [2]. In protein solutions of ionic strengths below 0.1 M, the interactions between protein molecules are dominated by the balance of Coulomb and van der Waals forces [66, 67, 84]. At physiological ionic strengths, the Debye screening length becomes 0.9–0.6 nm, i.e., comparable to the surface roughness of the protein molecules, see Fig. 2 [18, 66, 69], and the electrostatic repulsion becomes insignificant. Yet, protein molecules in solution do not spontaneously precipitate at physiological and higher ionic strengths. This mystery was resolved by the realization that in protein solutions intermolecular interactions are dominated by the structuring of water and other solvent molecules around the protein molecules [18, 60, 70, 71, 72]. Several theories have derived potentials of intermolecular interactions dominated by solvent structuring [18, 73, 74, 75]. While the effects of the solvent structuring on the protein phase diagrams have been explored in Monte Carlo simulations [76, 77], neither simulations nor theory addressing cluster formation in protein solutions exist.

For insight into the pair interactions between the lumazine synthase molecules, the dependence of the second osmotic virial coefficient B_2 on the concentration of the phosphate buffer was determined, Fig. 4. At all $C_{phosphate} < 1.4$ M, the values of B_2 are significantly greater than the values for non-interacting hard spheres, for which $B_2^{HS} = 2\pi N_A M_w^{-2} \int_0^{2R_1} r^2 dr = 4 V_m M_w^{-2}$ (R_1 is the radius of the molecule and $V_m = (4/3)\pi R_1^3 N_A$ is the molar volume of lumazine synthase), Fig. 2. The $B_2 > B_2^{HS}$ inequality suggests significant repulsion between the molecules. The Coulombic double layer forces [66, 67] are almost completely screened at ionic strengths >0.25 M [60, 78]. In a phosphate buffer at $pH = 8.7$, the phosphate ions are in

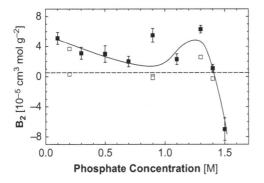

Fig. 4 The dependence of the second virial coefficient B_2 on the concentration of the phosphate buffer at 20°C. *Closed* and *open symbols* designate independent series of determinations. *Solid line* is just a guide for the eye. *Horizontal dashed line* indicates value of $B_2 = 3.2 \times 10^{-6}$ cm^3 mol g^{-2} for non-interacting hard spheres of volume and mass equal to those of the lumazine synthase molecule

divalent and trivalent states and the ionic strength is near 0.3 at $C_{phosphate} = 0.1$ M. Thus, the repulsion reflected in Fig. 4 is of non-electrostatic origin and is likely due to the structuring of the solvent: water, phosphate and cations, around the protein molecules. There is little understanding about the dynamics and in particular about the response of the structured solvent to the variations in the solution composition [60, 71, 72, 73, 74, 78]. While this makes the non-monotonic behavior of the data in Fig. 4 hard to interpret, there should be little doubt that the interactions are dominated by forces due to solvent structuring.

Interactions in hemoglobin solutions, in particular HbS, have been studied by osmometry [79, 80], sedimentation [81], as well as by light scattering [82, 83]. Since in the sequence HbA, S, and C (a mutant with a single mutation from glutamate to lysine at the $\beta 6$ site) the charge at the $\beta 6$ site changes from negative to neutral to positive, it has been speculated that this one elementary unit charge difference underlies, through electrostatic intermolecular interactions, the variability of the solid phases formed by these three variants. On the other hand, the deviations from ideality found by osmometry and sedimentation were solely attributed to the finite volume of the hemoglobin molecules [84, 85].

A recent determination of the second virial coefficient B_2 of four hemoglobin variants by static light scattering yielded a value fourfold higher than the value for non-interacting hard spheres [86]. These high B_2 values indicate strong repulsion between the Hb molecules, incompatible with a model of non-interacting hard spheres. Furthermore, it was found that the repulsion between the Hb molecules is not affected by the addition of electrolyte NaCl. It was also found that this effect is practically identical for all Hb variants tested, despite their charge differences. It was concluded that the observed repulsion is not of electrostatic origin. The weakness of the electrostatic forces was attributed to the chosen pH, which is very close to the isoelectric points for these Hb variants. This leads to small molecular charges, <5 [87]. On the other hand, it was found that the B_2 has different values in two tested buffers, phosphate and HEPES, suggesting that the strong intermolecular repulsion for all Hb mutants are due to the specific interaction with solvents components. Thus, similarly to lumazine synthase, the Hb molecules interact via forces due to the structuring of the solvent around the protein molecule.

4 Lack of Liquid–Liquid Phase Separation in Solutions of Lumazine Synthase and Hemoglobin

In search for an equilibrium dense liquid phase in solutions of lumazine synthase, the phase behavior of the solutions was probed under a broad range of conditions. Solution samples with protein concentration in the range 3–30 mg ml^{-1} and buffer concentration in the range 1.3–1.5 M were kept at temperatures ranging from 40°C (determined by the protein stability to denaturation) down to –4°C (below which the solution freezes) and monitored with differential interference contrast optics, sensitive to any heterogeneity larger than 0.5 μm. No evidence of formation of dense liquid droplets was found.

In similar experiments with oxy- and deoxy- hemoglobins A and S in 0.15 M phosphate buffer without any other additives, no liquid–liquid separation was observed even with protein concentration as high as 450 mg ml^{-1} [34]. This is in contrast to solutions containing as little as 0.1 or 0.25% polyethylene glycol 8000, in which L–L separation readily, reversibly, and reproducibly occurs immediately upon temperature increase [34].

To understand this lack of macroscopic dense liquid phase, we need to understand why these metastable phases are observable in solutions of other proteins. The dense liquids seen with the proteins lysozyme, γ-crystalline, with large organic molecules [34, 88, 89, 90, 91, 92, 93], and others, are metastable with respect to an ordered solid phase but, very importantly, are stable with respect to the initial low-density solution. After small droplets of such dense liquids form, their lifetime is limited by two events: decay into the initial solution or transformation into an ordered solid phase. The former is thermodynamically unfavorable, and the latter is extremely slow [94]. As a result, in these other systems the lifetime of the droplets is sufficient to allow growth to dimensions up to hundreds micrometers [95].

The lack of macroscopic dense liquid phases in solutions of lumazine synthase and the four Hb variants indicates that the region, in which such phase is stable with respect to the initial solution, is outside of the tested area in the (temperature, concentration) plane. We correlate this possibility with the intermolecular interaction potentials: recent Monte Carlo simulations have shown that if the intermolecular interaction potential possesses a maximum at intermolecular separations longer than those of the main attractive minimum, the metastable liquid–liquid separation is pushed to low temperatures [76]. A potential with a repulsive maximum is a natural assumption for lumazine synthase and hemoglobin – such potentials are an adequate representation of the water-structuring interactions, indicated by the B_2 data for both proteins [18, 96].

5 Dense Liquid Clusters in Solutions of Lumazine Synthase and Hemoglobin

5.1 Large Scatterers Revealed by Dynamic Light Scattering

Dynamic light scattering was employed to characterize solutions of lumazine synthase with different buffer concentrations and hemoglobin with a fixed buffer concentration. Data collection was initiated immediately after solution mixing and was repeated every 30 s for ~2–4 h. In all such runs with buffer concentration below 1.4 M, more than 80 or 90% of all collected correlation functions reveal the presence of a single scatterer with diffusivity $D = 1.5 \times 10^{-7}$ cm^2 s^{-1} [50]. Using Stokes law and the phosphate buffer viscosity [56], the size of these scatterers is $\sim$16.5 nm, consistent with the radius of hydrated lumazine synthase molecules. However, even among the first collected correlation functions we found those that deviate reveal the presence of a second process with a much slower decay rate: the correlation function $g_2(\tau)$ and the distribution function $G(\tau)$ were similar to those for hemoglobin in Fig. 3. If one assumes that the slow process is diffusion of large scatterers, one can evaluate their radii using the Stokes–Einstein relation, (4). The number density n_2 of the large scatterers and the fraction of the total solution volume φ_2, which they occupy are evaluated from (5) [44], with ρ_2 assumed to be $\sim$500 mg ml^{-1}, similar to dense liquids of other proteins [88, 89, 90, 91, 97, 98], and lower than the density of the lumazine synthase crystals.

For lumazine synthase, the amplitudes corresponding to the large species A_2, defined in Fig. 3, are from 2 to 4% of the amplitude for the lumazine synthase molecules A_1. Due to the errors inherent to the processing of light scattering data [53, 99], the determination of R_2 has an confidence interval of $\sim$30% of the determined value, while the relative error of the n_2 and φ_2 determination is up to 100% and they should be viewed as order of magnitude estimates.

The radii of the large scatterers R_2 are plotted in Fig. 5a as a function of the time after the beginning of dynamic light scattering data collection. The variations of the number densities and volume fractions occupied by the large scatterers are shown in Fig. 5b and c, respectively. Neither of the three plots in Fig. 5 show any trends toward lower or higher values. The frequency f of detection of clusters can be evaluated as the ratio of the number of data sets which reveal large scatterers – 25 – to the total number of data sets shown in Fig. 4, 241, so that $f = 0.10$.

Similar DLS data for deoxy-HbS, oxy-HbS, and oxy-HbA reveal that all three Hb variants, similar to lumazine synthase, exhibit a second process, significantly slower than the diffusion of single Hb molecules. Figure 3 illustrates these data for deoxy-HbS and shows that in contrast to lumazine synthase, the amplitudes A_2 corresponding to these slow processes are comparable to the amplitudes A_1 corresponding to single Hb molecules. This allows accurate characterization of the nature of the slow process.

In Fig. 6 we monitor the time evolution of the Hb cluster radii R_2. With all three Hb variants and at all concentrations probed, the clusters are present immediately after solution preparation and, in contrast to lumazine synthase, never disappear for the duration of data collection. The apparent fluctuations in the R_2's at $C_{deoxy-HbS} = 67$ mg ml^{-1}, $C_{oxy-HbS} = 108$ mg ml^{-1}, and $C_{oxy-HbA} = 51$ mg ml^{-1} are likely due to random noise in the dynamic light scattering signal [44]. Since the noise level decreases when the signal is stronger at higher Hb concentrations, the fluctuations in the traces with $C_{deoxy-HbS} = 131$ mg ml^{-1} and $C_{ocy-HbA} = 150$ mg ml^{-1} are weaker. Significantly, regardless of the fluctuations, the mean radii of the clusters of the three Hb variants at all probed concentrations do not increase over the 3 h of monitoring.

Fig. 5 Time evolution of lumazine synthase cluster characteristics: radius R_2, number density n_2, and volume fraction φ_2. $C = 8.1$ mg ml^{-1} in 1.3 M phosphate buffer with $pH = 8.7$ at 20°C. Correlation functions were collected every 30 s over 3 h. Correlation functions with a single peak, for lumazine synthase molecules, are not represented

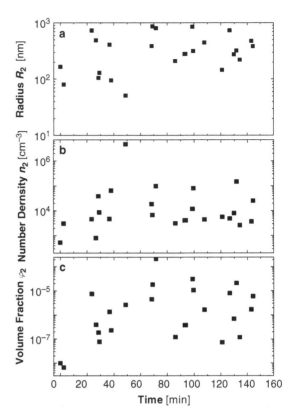

5.2 What the Large Scatterers Are Not

Tests with deionized water and low-concentration phosphate buffers showed lack of any scattering populations, while phosphate buffers at concentrations 0.5–1.3 M revealed scatterers of sizes of a few nanometers, much smaller than the lumazine synthase and the large scatterers; these are, likely, phosphate ion oligomers [100].

For tests of potential impurity effects, we added latex particles of 0.39 μm diameter to protein solutions, which exhibit time-dependent clustering. We found that if the volume fraction of the latex particles was $<10^{-8}$, they were undetectable in the correlation functions. However, whenever a particle passed through the illuminated solution volume, the scattered intensity spiked. This spike was averaged out in the correlation function. At particle volume fractions in the range 10^{-7}–10^{-5}, the correlation function appeared similar to the upper curve in Fig. 3a and the radius of the slow-decay time scatterers was $\sim$180–220 nm, corresponding to the introduced latex particles and different from the radius of the large scatterers. We conclude that the large scatterers reflected in Fig. 5 are not impurity particles.

The experiments with the added latex particles allow us to test another hypothesis for the origin of the slow process in the correlation function of the scattered

Fig. 6 Time dependence
of the radii of dense liquid
clusters in solutions of the
three hemoglobin variants as
indicated in the plots, at the
concentrations shown in the
plots at 25°C

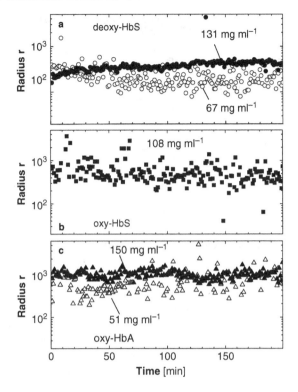

light. The slow decay rates may reflect not free Brownian motion of large scatterers, but diffusion of single protein molecules within immobile structures. This diffusion would be slow due to high effective viscosity. These structures would occupy part of the volume, so that single molecules in a solution of "normal" viscosity could still contribute to the signal; i.e., they could be either large clusters, greater than the several microns range detectable by lights scattering, or loose networks of molecules.

To exclude large dense clusters of protein molecules, we note that they would have been detected in our search for stable dense liquid phases. Loose networks of molecules could have low-refractive index and be optically undetectable. Such networks would increase the low-shear viscosity of the protein solution, but they would be destroyed by high-shear flow. The latter consideration allows tests for the presence of loose networks by determination of the low-shear viscosity of lumazine synthase solutions. However, the found low-shear viscosity of the oxy-HbA and deoxy-HbS solutions was in the range 2–4 cP, equal to the high-shear values in [57]. Analogously, the tests with lumazine synthase reproduced the published viscosity of the high-concentration phosphate buffers [56] (this viscosity is unaffected by the presence of the protein because of its low concentration). The equality of the low- and high-shear viscosity in solutions of lumazine synthase and hemoglobin reveals that no networks of molecules exist in these solutions and the large scatterers are in fact clusters of molecules of the respective protein.

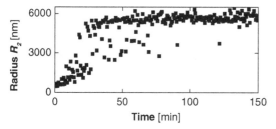

Fig. 7 Time evolution of aggregates during nucleation of crystals in lumasine synthase solutions. $C = 5.4$ mg ml^{-1} in 1.4 M phosphate buffer with $pH = 8.7$ at 20 °C. Almost immediately after solution mixing a population of large scatterers starts to grow and the autocorrelation function never goes back to a single exponential decay. After ~1.5 h numerous crystals are seen in the cuvette

If the concentration of the phosphate buffer in lumazine synthase solutions is increased above 1.4 M, the protein readily crystallizes. In such solutions, ~30 min after solution preparation a population of scatterers is superimposed on those similar to the one in Fig. 5 and in another ~15 min is the only one reflected in the light scattering signal, Fig. 7. After this population appears, the correlation function never goes back to a single scatters, and the radii and mass fraction of these scatterers grow indefinitely. After ~1.5 h, numerous tiny hexagonal lumazine synthase crystals are seen in the cuvette. We conclude that this growing population of scatterers consists of crystalline nuclei and supercritical crystallites of the protein. Conversely, the steady populations of scatterers in Figs. 5 and 6 are not crystals or other growing solid phases.

5.3 Mesoscopic Dense Liquid Clusters

Nucleation rates several orders of magnitude faster than those of crystals, and fast growth and decays rates have been shown for dense liquid droplets in protein solutions [95]. This and all other pieces of evidence discussed above lead to the conclusion that the large scatterers in solution of lumazine synthase in of the three tested Hb variants are clusters of a dense liquid phase.

Figure 5b and c shows that the concentration n_2 and the volume fraction φ_2 of the lumazine synthase clusters vary quite significantly. Part of the apparent variation may be due to the low accuracy of the determination, see above. On the average, the cluster concentration is $n_2 \approx 10^4$ cm^{-3}, and their fraction of the solution volume occupied by clusters is $\varphi_2 \approx 10^{-6}$. With hemoglobin S, the scatter of n_2 and φ_2 are lower, and the corresponding numbers are significantly higher, $n_2 \approx 10^{10}$ cm^{-3} and $\varphi_2 \approx 10^{-3}$ [44]. For oxy-HbA the average $n_2 \approx 10^7$ cm^{-3} and $\varphi_2 \approx 10^{-4}$ [44]. Comparisons of the n_2 and φ_2 values of these three proteins suggest that the cluster concentration is governed by the concentration of protein molecules, much lower in the tested solution of lumazine synthase than of the two Hb variants, as well as by the intermolecular attraction.

The disappearance of the lumazine synthase clusters from the light scattering signal in ~90% of the data sets in Fig. 5 is surprising: the cluster population in solutions of the three hemoglobin variants in Fig. 6 is steady throughout the entire duration of solution monitoring. This time dependence could mean one of two possibilities: the lumazine synthase clusters appear and completely disappear and then reappear again, or the sizes and number of clusters in the scattering population vary around a detection threshold. To decide between these, we analyze in Fig. 8a the distribution of times between detection of clusters: they vary from 30 s to 17 min, with a mean of ~3 min. In this distribution, longer times are less likely and this suggests that a population of clusters, with varying clusters' sizes R_2 and number n_2, exists at all times. As revealed by the tests with the latex particles, the population of large scatterers is only detectable if its $\varphi_2 > 10^{-8}$.

The conclusion of large undetectable population of lumazine synthase clusters was tested by experiments with varying protein concentration, from 1.3 to 8.1 mg ml^{-1} [101]. They revealed that at all tested protein concentrations, the sizes of the clusters R_2 within the detectable populations are randomly distributed, as in Fig. 8b, with a mean $\bar{R}_2$ ~350 nm. The number density n_2 of the clusters and the volume fraction φ_2 which they occupy do not increase with increasing protein concentration either. The only variable, which increases, is the frequency f of detection of the clusters: from 0.045 to 0.11 of all data sets. This observation supports the above conclusion that a population of clusters exists at all times and is detected only when

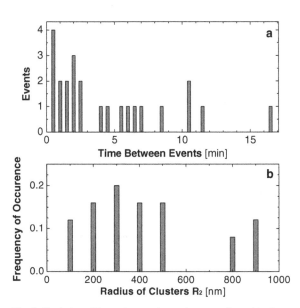

Fig. 8 Statistics of lumazine synthase clusters reflected in time trace in Fig. 5a. (**a**) Distribution of the time intervals between detection of clusters. (**b**) Distribution of clusters' sizes

the cluster volume fraction overcomes the detection threshold. The reproducibility of the mean cluster size suggests that this is a preferred size, at which a functional, which governs cluster formation, has a local minimum. The wide distribution of the cluster sizes suggests that this minimum is wide and shallow.

The dependencies of the cluster radii R_2 and volume fractions φ_2 on the Hb concentration are not identical for the three Hb variants, Fig. 9. For deoxy HbS and oxy-HbA both R_2 and φ_2 increase as Hb concentration is increased. This seems to be the expected behavior—higher Hb concentration in the solution leads to stronger driving force for cluster formation. Quite surprisingly, the clusters for oxy-HbS decrease in size and volume fraction and at $C_{oxy-HbS} > 150$ mg ml^{-1} no clusters are observed. Currently, no explanation of these dependencies can be offered.

5.4 The Cluster Lifetimes and Stability with Respect to the Solution

The lumazine synthase clusters are detected only if a sufficiently large population exists for a time comparable to the time of collection of the correlation function, 30 s. On the other hand, Fig. 5 shows few instances of neighboring moments with large scatters. These two observations provide a low and a high limit of the time that a cluster population is detectable and set it of order 10 s.

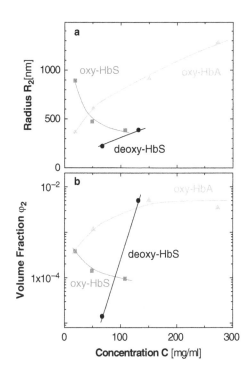

Fig. 9 The dependencies of the radii of the dense liquid clusters R_2 and the fractions of the total solution volume φ_2 occupied by them on the concentration of each respective Hb variant as indicated in the plots at 25°C. Each data point in the average of a trace similar to those in Fig. 6. Error bars represent intervals of 95% confidence of these averages and are comparable to symbol size in both a and b. Solid lines are just guides for the eye. No clusters of oxy-HbS are detected at this temperature in solutions of concentration higher than 100 mg ml^{-1}

The evaluation of the lifetime of the HbS clusters is more difficult: the data in Fig. 6 do not allow it. In fact, if the clusters have a limited lifetime, it is possible that the slow process reflected by the correlation functions in Fig. 3 is not the diffusion of the clusters, but their decay. These two processes can be distinguished if the dependence of the characteristic rate of the slow process Γ_2 on the scattering vector $\mathbf{q}$ is available: according to [102]

$$\Gamma_2 = \Gamma_0 + D_2\mathbf{q}^2, \tag{10}$$

where Γ_0 is the rate of cluster decay, whose contribution to Γ_2 does not depend on the scattering vector $\mathbf{q}$, and D_2 is the cluster diffusion coefficient.

The two limiting cases of (17) are $\Gamma_0 \gg D_2\mathbf{q}^2$, and $\Gamma_0 \ll D_2\mathbf{q}^2$. In the first case $\Gamma_2 = \Gamma_0$, Γ_2 does not depend on the scattering vector, i.e., on the scattering angle, and the characteristic time of the slow process in Fig. 3 is in fact the cluster lifetime. In the second case, $\Gamma_2 = D_2q^2$, the slow process is the cluster diffusion, and the cluster lifetime is longer than the characteristic diffusion time τ_y. The strong absorbance of Hb solutions at all visible wavelengths prevents the analyses of the angular dependence of Γ_2. Approximate determinations at scattering angles of 30, 60, and 90° revealed that Γ_2 is a strong function of the scattering vector. Thus, determinations of the cluster sizes R_2 using (4) and assuming that $\Gamma_2 = D_2q^2$ do not carry a significant bias. From the inequality $\Gamma_0 \ll D_2q^2$ with $q^2 = 3.5 \times 10^{10}$ cm^{-2} and $D_2 = 2 \times 10^{-9}$ cm^{2}s^{-1}, $\Gamma_0 \ll 70$ s^{-1}, and the HbS cluster lifetimes have a lower bound $1/\Gamma_0 \approx 15$ ms.

The short lifetimes of these clusters indicate that they have a free energy higher than not only that of the crystals, but also of the low-concentration solution. On the other hand, their measurable lifetime suggests that a barrier, either thermodynamic or kinetic, for the droplets' decay exists, i.e., they are metastable and not unstable with respect to the initial state. In terms of the free energy landscape in Fig. 1, the clusters represent metastable phases that occupy relatively shallow free energy minima [103, 104].

5.5 Direct Imaging of Lumazine Synthase Clusters and Their Stability with Respect to Crystals

In solutions of lumazine synthase, there is a single stable condensed phase: the hexagonal crystals. To probe the stability of the dense liquid clusters with respect to the crystals, we monitor the growth of (001) faces of lumazine synthase crystals [42].

Figure 10 is one of several which show clusters, which land on the crystal surface. While the lateral dimensions of the clusters cannot be judged from the images, their height is reliably determined: it varies between ∼100 and several hundred nanometers, in Fig. 10a the size is ∼120 nm. The clusters do not decay because of their interaction with the crystal, and grow sideways and become integral parts of the crystal, generating, in Fig. 10b, five new layers.

Fig. 10 Sedimentation of a 3D-object in (**a**) and its development into a stack five crystalline layers in (**b**) in a lumazine synthase solution of 3 mg ml^{-1} in 1.3 M phosphate buffer at pH = 8.7. Tapping mode AFM imaging, scan size 20 × 20 mm, time interval between a and b 9 min. (**c**) and (**d**) height profiles along a *horizontal line* crossing 3D-object in (a) and (b), respectively, show the object height of ∼120 nm immediately after sedimentation in (a) and ∼75 nm in (b). *Arrows* in (a), (b), (c) and (d) mark same crystal layer. From [42]

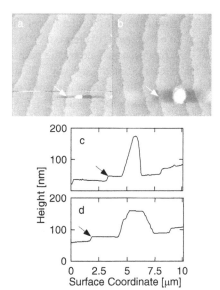

Figure 10 demonstrates that these are clusters of dense liquid: (i) The clusters shrink in height as they rest on the crystal surface, compare Fig. 10c and d. (ii) The layers originating from the clusters merge continuously with each other and with the underlying lattice. (iii) The velocity of the layers originating from a cluster is the same as the velocity of the other layers on the surface of the crystal: compare locations of two types of layers in Fig. 10a and b. If these were crystalline clusters, the probability of them landing with a (001) plane downward, and rotating to a perfect registry with the underlying lattice would be negligible. Numerous examples of microcrystals landing on surfaces of growing crystals out of registry and getting incorporated with major stacking defects have been reported [105, 106]. Disordered solid clusters would not shrink in size within the observation times, and the generation of new layers would likely be accompanied by the creation of a strained shell that would delay the spreading of the layers started by the clusters [106].

The participation of the metastable dense liquid clusters in the growth process illustrated by Fig. 10 can be understood in the following way. The metastable liquid phase clusters have high molecular density, comparable to that in a solid state, and high mobility of the protein molecules. The encounter of a liquid cluster with the crystal surface helps the dense liquid to overcome the barrier toward the deeper free energy minimum: under the influence of the periodic field of the crystal, long-range order is imposed in the dense liquid and it transforms into stacks of crystalline layers.

An interesting feature of the dense liquid clusters is that at supersaturations $C/C_e < 2$, they are crucial for the growth of the crystals. At supersaturations $C/C_e > 2$, the growth layers are generated by 2D-nucleation, with numerous circular islands seen on the surface of the crystals [107, 108, 109, 110, 111]. This mode of layer generation requires a driving force above a threshold [112]. No two-dimensional nucleation was observed at $C/C_e < 2$. Furthermore, no dislocations, an alternative source of layers, operational even at low supersaturations [107, 113], were seen outcropping on faces of lumazine synthase crystals during the AFM observations. However, the crystals grow even at driving forces as low as $C/C_e \approx 0.5$ only because new layers originate from clusters landing on the crystal surface, as in Fig. 10.

5.6 HbS Clusters and Nucleation of Sickle Cell Hemoglobin Polymers

It has been shown that the nucleation of deoxy-HbS polymers proceeds via a two-step mechanism, i.e., the nuclei of the polymer fibers formed inside the clusters discussed here [114]. This mechanism is illustrated in Fig. 11. While clusters have not been reported for solution of proteins other than those discussed here, there is significant evidence that this "two-step mechanism" applies to many proteins, as well as to colloid, and small-molecule organic and ionic substances [32, 33, 34, 35, 36, 37, 38, 39, 40]. Even more importantly, the results in [114] show that the rate of nucleation of the HbS polymers is limited by the rate of formation of the clusters. This is because in contrast to nucleation of crystals, the nucleation of HbS polymers

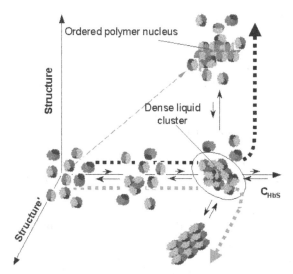

Fig. 11 Schematic representation of the two-step nucleation mechanism of HbS polymers. The nucleation pathway in the space of order parameters. Horizontal axis is HbS concentration, along which dilute solution and dense liquid can be distinguished. Axes of evolution of ordered structures are orthogonal to concentration axis, only two leading to polymers and crystals, respectively, are shown. Other possible structure axes include other crystal polymorphs, disordered aggregates, and gels. Thin dashed arrow along diagonal indicates pathway of the one-step nucleation mechanism of HbS polymers. Thick short-dashed lines indicate pathways of the two-step mechanism leading to polymers or crystals, respectively

is very fast and occurs over time scales of a few seconds. Thus, by controlling the stability and the rate of formation of the clusters, one can control the rate of HbS polymer nucleation.

The results with addition of PEG and increase of phosphate concentration [34, 92, 115, 116] show that the stability of the dense liquid is strongly affected by the solution composition. The nature of additives that affect the stability of the dense liquid requires detailed further studies. Still, these considerations lead to a very significant conclusion: the presence of clusters and their role in the HbS polymer nucleation provide a new handle for control of the nucleation process. The previous outlook on HbS polymer nucleation via a direct classical-type mechanism [16] allows for a single control parameter at fixed temperature: the deoxy-HbS activity. The realization that other solution components may have strong effects on the nucleation rate, and that these effects would occur through the volume fraction occupied by dense liquid clusters offers two novel insights: (i) that much of the variability of the clinical manifestations of sickle cell anemia may be due to the numerous non-protein components of the red blood cell cytosol, and (ii) that a treatment could be sought among these components or among other compounds which would penetrate the red cells and delay the polymerization kinetics by lowering the cluster concentration.

5.7 Why Do Clusters Exists?

The puzzle opened by the clusters is related to an apparent contradiction: for the clusters to exist, the molecules should attract; if the molecules attract, the clusters should grow to macroscopic dimensions. Thus, a microscopic theory of mesoscopic clusters should provide a mechanism of increase of the clusters' free energy as they grow in size. This extra free energy makes them metastable and, when it reaches the magnitude of the gain due to molecular attraction, determines the cluster size.

The existing microscopic models of cluster formation explain their finite size with the balance of intermolecular attraction, due to van der Walls, hydrophobic or other forces, and repulsion between like-charged species [8]. The main result of this theory is that large clusters could be expected only if the electrostatic repulsive energy is high. These clusters of more than 100 molecules require repulsive energies in the order of 10^4 $k_B T$ [8]. This theory has been found to adequately describe the existence of clusters of a few tens of particles or molecules in colloid suspensions [11, 117, 118] and in solutions of the protein lysozyme at low ionic strength [11].

This mechanism is inapplicable to the clusters of lumazine synthase because even at the lowest phosphate concentration tested, 0.1 M, the ionic strength of nearly 0.3 M precludes the action of any electrostatic forces. Moreover, if one thinks of intermediate range repulsion due to solvent structuring as a replacement to electrostatic repulsion, one still fails to explain cluster sizes. The reason is that forces due to solvent structuring only act on immediate neighbors: they cannot penetrate these neighbors and act on a third molecule.

The electrostatic theory is inapplicable to the clusters of the hemoglobin variants either: with sizes of several hundred nanometers, they likely contain order of 10^6 molecules. If these clusters form according to the mechanism put forth in [8], they would require a strong electrostatic repulsion, which is hard to envision for the almost neutral Hb molecules under near-physiological pH. Indeed the isoionic pH for HbA is 6.8 [43, p. 134] and slightly higher for HbS, and the R-to-T transition has a minor effect on this value. Thus at the working $pH = 7.35$, the Hb molecules carry around one negative charge. If the Hb molecules attain charge after they enter the clusters, the high level of electrostatic energy required for clusters as large as those observed would drive apart different parts of the Hb molecule and denature the molecules (V. Lubchenko, private communication, 2006).

An even more general statement is that the microscopic mechanism of formation of clusters, as large as those with lumazine synthase, cannot be based on the interactions between pairs of molecules or particles. Indeed, a cluster with $R_2 = 350$ nm has 15–20 molecules along its diameter. All interactions in protein solutions, other than electrostatic, have ranges shorter than 1/3 of the molecular size [18]; thus, none of them can act on lengths of several molecular sizes.

6 Monte Carlo Simulations of Formation and Decay of Clusters

For additional insight into the mechanism leading to cluster formation, we carried out simple Monte Carlo simulations. Protein molecules were modeled as spheres interacting with short-range potentials with a single extremum, a minimum, or with an additional longer range repulsive maximum. We used potentials introduced in [76, 77]

$$U(r) = \begin{cases} \infty & r \leq \sigma \\ \dfrac{4\varepsilon}{\alpha^2} \left\{ \dfrac{1+h(r)}{\left[(r/\sigma)^2 - 1\right]^6} - \dfrac{\alpha}{\left[(r/\sigma)^2 - 1\right]^3} \right\} & \sigma < r \end{cases}, \qquad (11)$$

where ε is the depth of the attractive minimum, σ is the molecular diameter, α determines the width of attraction [77], and the "hump" function

$$h(r) = \begin{cases} 0 & r \leq r_{st} \\ \gamma[(r/\sigma)^2 - (r_{st}/\sigma)^2]^2 & r > r_{st} \end{cases}, \qquad (12)$$

with docking point of "hump" r_{st} and effective "hump" height γ represents an intermediate range repulsion for $\gamma > 0$. Potentials with varying $h(r)$, including $h(r) = 0$, were found to adequately reproduce several features of the equilibria between protein solutions and crystals and between solutions and dense liquids [76, 77].

Monte Carlo simulations of the evolution of an initially homogenous system were performed at constant temperature T, $k_B T/\varepsilon = 0.48$, where k_B is the Boltzmann constant and ε is the depth of the attractive minimum in the intermolecular potential, with a fixed number of particles, $N = 2000$, and volume $V = 10,000$. The duration of the simulation was 200000 cycles, where each cycle consists of 2000, i.e., one per each particle, displacement attempts for randomly selected particles. The displacement was limited to 0.3 of the particle diameter [76, 77].

For simulations of cluster formation, potentials with $\alpha = 50$ as in [32], $r_{st} = 1.16$, and $\gamma = 0$, 100, 350, or 520, Fig. 12d, were used. The liquidus and solidus lines of molecules interacting with these potentials are shown in Fig. 12b [76]. The liquid–liquid coexistence line for $\gamma = 0$ reaches its upper critical point at $T = 0.40\varepsilon/k_B$, all other liquid–liquid coexistence lines are at even lower temperatures [76, 77]. Simulations were carried out at volume fraction of molecules 0.14, at which for the chosen temperature all four studied systems are just barely supersaturated with respect to the crystals. Because of this low supersaturation, crystals never form in the simulations.

Clusters are defined as groups of molecules forming a continuous series of contacts. Two molecules form a contact if the distance between their centers is $<1.23\sigma$, as in [35]. Aggregations of less than 8 particles are not considered as clusters, and this limit defines the birth and the death of a cluster; in some cases cluster formation/dissolution is the result of coalescing/breaking of two or more aggregates. Between its birth and death, a cluster's size increases and decreases several times. Typically, several clusters coexist in the simulation volume. All clusters are liquid: local bond order analyses [119, 120] revealed that they were disordered [76] and monitoring of the relative positions of molecules within clusters revealed that they flow.

The first significant result of the simulations is that clusters exist with all studied potentials. All clusters have limited lifetimes, which are significantly shorter than the total length of the simulation runs. The cluster lifetime is quantified as the number of Monte Carlo steps between its birth and death, the cluster's size is defined as the maximum number i of molecules contained in it during its lifetime. For each size i, an average lifetime is calculated from all clusters in the simulations, and the results are plotted in Fig. 12a.

Clusters exist even in simulations with a single attractive minimum without hump, Fig. 12a and b. This observation supports the conclusion, formulated above, that the existence of large clusters is not determined by the interactions between pairs of molecules. Note that the addition of intermediate-range repulsion increases the clusters' lifetimes and highlights the role of solvent structuring for the clusters' existence.

The clusters' stability is evaluated as the natural logarithm of the probability of having a cluster of a size i, P_i/P_1, where P_i is the number of clusters of size i, and P_1 is the number of single molecules in the system. According to [32], $\ln(P_i/P_1)$ is proportional to the chemical potential $\Delta\mu_i$ of a cluster of a certain size i. This $\Delta\mu_i$ is

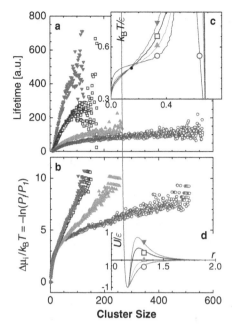

Fig. 12 Monte Carlo simulations of cluster formation. (**a**) Increase of cluster lifetime and (**b**) increase of chemical potential $\Delta\mu$ of clusters with cluster size i for four potentials shown in (**d**). (**c**) (Volume Fraction φ, Temperature T) phase diagrams corresponding to the four potentials used, temperature is scaled with potential depth ε/k_B, k_B – Boltzmann constant. Black dot denotes φ and T for simulations in (**a**) and (**b**). (**d**) Intermolecular pair potentials $U(r)$ used in simulations whose results are shown in a–c. Interaction energy U is scaled with potential depth ε. Symbols in (**c**) and (**d**) correspond to symbols in (**a**) and (**b**)

not to be confused with the chemical potential of the protein within the cluster—the latter is poorly defined because of the small clusters' sizes and the related significant surface effects.

The plots in Fig. 12a and b show that both the clusters' lifetimes and free energy increase as their size i increases. The increase of the clusters' free energy is expected – for the studied systems, the dense liquid is unstable with respect to the dilute state and may become stable only at significantly lower temperatures [76]. The correlation between higher free energy and longer lifetimes is supported by comparisons between results with different potentials in the same Fig. 12a and b. This correlation indicates that the clusters' lifetimes are not determined by their excess free energy – if this were the case, higher free energy would lead to greater driving force for decay and faster decay. We conclude that clusters' lifetimes are determined by the kinetics of their growth and dissolution. One factor for slower kinetics is the height of the hump, which prevents molecules from detaching from their neighbors; another factor may be the density and viscosity of the dense liquid inside the clusters.

At larger cluster sizes i, at which fewer clusters were sampled for the averages in Fig. 12a and b, the data scatter is greater. This allows the observation that some populations of clusters of a certain size have lower free energy than populations of smaller sizes, likely because of different structures and densities. This variability of the clusters' stability explains the wide variability of clusters' sizes seen in Figs. 4, 6, and 7.

7 Summary and Perspectives for Future Work

Using a combination of experimental methods, we demonstrate a novel feature of protein solutions and, likely, of solution of other large molecules and colloid-size particles: the presence of dense liquid clusters, which are metastable not only with respect to a solid phase, but also with respect to the low-concentration protein solution. These clusters play a crucial role for the growth of ordered solids in protein solutions. It is likely that similar clusters form in the crowded, high-protein environment of the cytosol of living cells, where they affect solution viscosity and the rate of all transport and biochemical processes.

The clusters of the protein lumazine synthase have characteristic lifetimes, after which they decay into the solution, of the order of 10 s. Only the lower bound of the clusters of three hemoglobin variants could be estimated as 15 ms. The sizes and fractions of solution volume occupied by the clusters vary in time. At high-protein concentrations, the clusters occupy from 10^{-5} to 10^{-3} of the solution volume and are always detectable. At low-protein concentrations, the clusters occupy up to 10^{-5} of the solution volume and are detectable only if this volume fraction exceeds 10^{-8}.

Using a simple Monte Carlo simulation scheme with protein-like potentials, we reproduce several essential features of the clusters: their metastability, their limited lifetimes, and the broad variability of their sizes. The simulations revealed that structuring of the solvent around the protein molecules contributes to the clusters lifetimes, but is not essential for their existence.

At this point, there exists no microscopic theory that could explain at the molecular level the mechanism of cluster stability, and the factors which determine their lifetimes and sizes. The properties of the clusters discussed here can serve as the basis for the formulation of such a theory. Particularly suggestive is the observation that while the sizes and volume fractions of the deoxy-HbS and oxy-HbA clusters increase at higher hemoglobin concentrations, those characteristics for oxy-HbS decrease. The counterintuitive behavior of oxy-HbS suggests that the cluster mechanism cannot be understood by focusing on the intermolecular interactions in the solution. It is likely that the disappearance of the oxy-HbS clusters at higher Hb concentrations is due to a change of the cluster properties, which lowers the cluster stability.

Acknowledgements We thank V. Lubchenko and A. Kolomeisky for numerous helpful discussions on the nature of the metastable phases; N. Neumaier, S. Weinkauf, Ilka Haase, Markus Fischer, and Adelbert Bacher for providing lumazine synthase and for participation in the early light scattering experiments; I. Reviakine for help with AFM imaging; and M. Shah for experimental support and suggestions. This work was supported by the National Heart, Lung, and Blood Institute, NIH, the Office of Physical and Biological Research, NASA, and by the University of Houston.

References

1. K. Luby-Phelps, Cytoarchitecture and physical properties of cytoplasm: Volume, viscosity, diffusion, intracellular surface area, *Int. Rev. Cytol.* **192**, 189 (2000).
2. R. J. Ellis, Macromolecular crowding: An important but neglected aspect of the intracellular environment, *Cur. Opin. Struct. Biol.* **11**, 114–119 (2001).
3. G. H. Pollack, *Cells, Gells, and the Engines of Life*. Ebner and Sons, Seattle, 2001.
4. H. Walter, Consequences of phase separation in cytoplasm, *Int. Rev. Cytol.* **192**, 331 (2000).
5. J. W. Gibbs, On the equilibrium of heterogeneous substances, *Trans. Connect. Acad. Sci.* **3**, 108 (1876).
6. J. W. Gibbs, On the equilibrium of heterogeneous substances, *Trans. Connect. Acad. Sci.* **16**, 343 (1878).
7. J. D. van der Waals, The equation of state for gases and liquids, in *Nobel Lectures, Physics 1901–1921*. Elsevier Publishing Company, Amsterdam, 1910.
8. F. Sciortino, S. Mossa, E. Zaccarelli, and P. Tartaglia, Equilibrium cluster phases and low-density arrested disordered states: The role of short-range attraction and long-range repulsion, *Phys. Rev. Lett.* **93**, 055701 (2004).
9. J. Groenewold and W. K. Kegel, Anomalously large equilibrium clusters of colloids, *J. Phys. Chem. B* **105**, 11702 (2001).
10. R. P. Sear, S.-W. Chung, G. Markovich, W. M. Gelbart, and J. R. Heath, Spontaneous patterning of quantum dots at the air-water interface, *Phys. Rev. E* **59**, R6255 (1999).
11. A. Stradner, H. Sedgwick, F. Cardinaux, W. C. K. Poon, S. U. Egelhaaf, and P. Schurtenberger, Equilibrium cluster formation in concentrated protein solutions and colloids, *Nature* **432**, 492 (2004).
12. G. B. Benedek, J. Pande, G. M. Thurston, and J. I. Clark, Theoretical and experimental basis for the inhibition of cataract, *Prog. Retin. Eye. Res.* **18**, 391 (1999).
13. M. Bucciantini, E. Giannoni, F. Chiti, F. Baroni, L. Formigli, J. Zurdo, N. Taddei, G. Ramponi, C. M. Dobson, and M. Stefani, Inherent toxicity of aggregates implies a common mechanism for protein misfolding diseases, *Nature* **416**, 507 (2002).
14. R. W. Carrell and D. A. Lomas, Conformational diseases, *Lancet* **350**, 134 (1997).
15. G. Dodson and D. Steiner, The role of assembly in insulin's biosynthesis, *Curr. Opin. Struct. Biol.* **8**, 189 (1998).
16. W. A. Eaton and J. Hofrichter, Sickle cell hemoglobin polymerization, in *Advances in Protein Chemistry* (C. B. Anfinsen, J. T. Edsal, F. M. Richards, and D. S. Eisenberg, eds.), Vol. 40, p. 63. Academic Press, San Diego, 1990.
17. C. M. Dobson, Getting out of shape, *Nature* **418**, 729 (2002).
18. P. G. Vekilov and A. A. Chernov, The physics of protein crystallization, in *Solid State Physics* (H. Ehrenreich and F. Spaepen, eds.), Vol. 57, p. 1. Academic Press, New York, 2002.
19. M. J. Stuart and R. L. Nagel, Sickle-cell disease, *Lancet* **364**, 1343 (2004).
20. G. R. Sergeant, Sickle-cell disease, *Lancet* **350**, 725 (1997).
21. V. M. Ingram, A specific chemical difference between the globins of normal human and sickle cell anaemia haemoglobin, *Nature* **178**, 792 (1956).
22. L. Pauling, H. A. Itano, S. J. Singer, and I. C. Wells, Sickle cell anemia, a molecular disease, *Science* **111**, 543 (1949).
23. R. Pawliuk, K. A. Westerman, M. E. Fabry, E. Payen, R. Tighe, E. E. Bouhassira, S. A. Acharya, J. Ellis, I. M. London, C. J. Eaves, R. K. Humphries, Y. Beuzard, R. L. Nagel, and P. Leboulch, Correction of sickle cell disease in transgenic mouse models by gene therapy, *Science* **294**, 2368 (2001).

24. E. V. Hahn and E. B. Gillespie, Sickle cell anemia: Report of a case greatly improved by splenectomy. Experimental study of sickle cell formation, *Arch. Int. Med.* **39**, 233 (1927).

25. H. Hofrichter, P. D. Ross, and W. A. Eaton, Kinetics and mechanism of deoxyhemoglobin S gelation: A new approach to understanding sickle cell disease, *Proc. Natl. Acad. Sci. USA* **71**, 4864 (1974).

26. A. S. Mehanna, Sickle cell anemia and antisickling agents then and now, *Curr. Med. Chem.* **8**, 79 (2001).

27. E. Vichinsky, New therapies in sickle cell disease, *Lancet* **360**, 4350 (2002).

28. S. T. Ohnishi, Introduction, in *Membrane Abnormalities in Sickle Cell Disease* (S. T. Ohnishi and T. Ohnishi, eds.), CRC Press, Boca Raton, 1994.

29. F. A. Ferrone, H. Hofrichter, and W. A. Eaton, Kinetics of sickle cell hemoglobin polymerization I. Studies using temperature jump and laser photolysis techniques, *J. Mol. Biol.* **183**, 591 (1985).

30. F. A. Ferrone, H. Hofrichter, and W. A. Eaton, Kinetics of sickle cell hemoglobin polymerization. II. A double nucleation mechanism, *J. Mol. Biol.* **183**, 611 (1985).

31. R. E. Samuel, A. E. Guzman, and R. W. Briehl, Kinetics of hemoglobin polymerization and gelation under shear: II. The joint concentration and shear dependence of kinetics, *Blood* **82**, 3474 (1993).

32. P. R. ten Wolde and D. Frenkel, Enhancement of protein crystal nucleation by critical density fluctuations, *Science* **277**, 1975 (1997).

33. O. Galkin and P. G. Vekilov, Control of protein crystal nucleation around the metastable liquid-liquid phase boundary, *Proc. Natl. Acad. Sci. USA* **97**, 6277 (2000).

34. O. Galkin, K. Chen, R. L. Nagel, R. E. Hirsch, and P. G. Vekilov, Liquid-liquid separation in solutions of normal and sickle cell hemoglobin, *Proc. Natl. Acad. Sci. USA* **99**, 8479 (2002).

35. A. Lomakin, N. Asherie, and G. B. Benedek, Liquid-solid transition in nuclei of protein crystals, *Proc. Natl. Acad. Sci. USA* **100**, 10254 (2003).

36. P. G. Vekilov, Dense liquid precursor for the nucleation of ordered solid phases from solution, *Cryst. Growth Des.* **4**, 671 (2004).

37. B. Garetz, J. Matic, and A. Myerson, Polarization switching of crystal structure in the nonphotochemical light-induced nucleation of supersaturated aqueous glycine solutions, *Phys. Rev. Lett.* **89**, 175501 (2002).

38. D. W. Oxtoby, Crystals in a flash, *Nature* **420**, 277 (2002).

39. W. Pan, A. B. Kolomeisky, and P. G. Vekilov, Nucleation of ordered solid phases of protein via a disordered high-density state: Phenomenological approach, *J. Chem. Phys.* **122**, 174905 (2005).

40. D. Kashchiev, P. G. Vekilov, and A. B. Kolomeisky, Kinetics of two-step nucleation of crystals, *J. Chem. Phys.* **122**, 244706 (2005).

41. H. Frauenfelder, S. Sligar, and P. Wolynes, The energy landscapes and motions of proteins, *Science* **254**, 1598 (1991).

42. O. Gliko, N. Neumaier, W. Pan, I. Haase, M. Fischer, A. Bacher, S. Weinkauf, and P. G. Vekilov, A metastable prerequisite for the growth of lumazine synthase crystals, *J. Am. Chem. Soc.* **127**, 3433 (2005).

43. D. L. Nelson and M. M. Cox, *Lehninger's Principles of Biochemistry*, third edition. W. H. Freeman, New York, 2000.

44. W. Pan, O. Galkin, L. Filobelo, R. L. Nagel, and P. G. Vekilov, Metastable mesoscopic clusters in solutions of sickle cell hemoglobin, *Biophys. J.* **92**, 267 (2007).

45. K. Schott, R. Ladenstein, A. König, and A. Bacher, The lumazine synthase/riboflavin synthase complex from *Bacillus subtilis*: Crystallization of reconstituted icosahedral β_{60} capsids., *J. Biol. Chem.* **265**, 12686 (1990).

46. K. Ritsert, R. Huber, D. Turk, R. Ladenstein, K. Schmidt-Bäse, and A. Bacher, Studies of the lumazine synthase/riboflavin synthase complex of *Bacillus subtilis:* Crystal structure analysis of reconstituted, icosahedral β-subunit capsids with bound substrate analogue inhibitor at 2.4 A resolution, *J. Mol. Biol.* **253**, 151 (1995).

47. A. Bacher, R. Baur, U. Eggers, H. Harders, M. K. Otto, and H. Schnepple, Riboflavin synthases of Bacillus subtilis, *J. Biol. Chem.* **255**, 632 (1980).

48. A. Bacher, S. Eberhardt, M. Fischer, K. Kis, and G. Richter, Biosynthesis of vitamin B₂ (riboflavin), *Annu. Rev. Nutr.* **20**, 153 (2000).
49. M. F. Perutz, Structure and mechanism of haemoglobin, *Br. Med. Bull.* **32**, 195 (1976).
50. K. S. Schmitz, *Dynamic Light Scattering by Macromolecules*. Academic Press, New York, 1990.
51. D. Eisenberg and D. Crothers, *Physical Chemistry with Applications to Life Sciences*. The Benjamin/Cummins, Menlo Park, 1979.
52. P. S. Berry, S. A. Rice, and J. Ross, *Physical Chemistry*. Oxford University Press, New York, 2000.
53. K. Schatzel, Single-photon correlation techniques, in *Dynamic Light Scattering. The Method and Some Applications* (W. Brown, ed.), p. 76. Clarendon Press, Oxford, 1993.
54. S. W. Provencher, CONTIN: A general purpose constrained regularization program for inverting noisy linear algebraic and integral equations, *Comp. Phys. Commun.* **27**, 229 (1982).
55. S. W. Provencher, A constrained regularization method for inverting data represented by linear algebraic equations, *Comp. Phys. Commun.* **27**, 213 (1982).
56. *CRC Handbook on Chemistry and Physics*. CRC Press, LLC, 2004.
57. P. D. Ross and A. P. Minton, Hard quasi-spherical model for the viscosity of hemoglobin solutions, *Biochem. Biophys. Res. Commun.* **76**, 971 (1977).
58. P. Stepanek, Data analysis in dynamic light scattering, in *Dynamic Light Scattering. The Method and Some Applications* (W. Brown, ed.), p. 177. Clarendon Press, Oxford, 1993.
59. B. H. Zimm, The scattering of light and the radial distribution function of high polymer solutions, *J. Chem. Phys* **6**, 1093 (1948).
60. D. N. Petsev, B. R. Thomas, S.-T. Yau, and P. G. Vekilov, Interactions and aggregation of apoferritin molecules in solution: Effects of added electrolytes, *Biophys. J.* **78**, 2060 (2000).
61. *Handbook of Chemistry and Physics,* 81th Edition. CRC Press, Boca Raton, 2000–2001.
62. I. S. Sohn and R. Rajagopalana, Microrheology of model quasi-hard-sphere dispersions, *J. Rheol.* **48**, 117 (2004).
63. I. Reviakine, W. Bergsma-Schutter, and A. Brisson, Growth of protein 2-d crystals on supported planar lipid bilayers imaged in situ by AFM, *J. Struct. Biol.* **121**, 356 (1998).
64. S.-T. Yau, D. N. Petsev, B. R. Thomas, and P. G. Vekilov, Molecular-level thermodynamic and kinetic parameters for the self-assembly of apoferritin molecules into crystals., *J. Mol. Biol.* **303**, 667 (2000).
65. A. J. Malkin and A. McPherson, Probing of crystal interfaces and the structures and dynamic properties of large macromolecular ensembles with in situ atomic force microscopy, in *From Fluid-Solid Interfaces to Nanostructural Engineering. vol. 2. Assembly in Hybrid and Biological Systems* (J. J. De Yoreo and X. Y. Lui, eds.), p. 201. Plenum/Kluwer Academic, New York, 2004.
66. M. Muschol and F. Rosenberger, Interaction in undersaturated and supersaturated lysozyme solutions: Static and dynamic light scattering results, *J. Chem. Phys.* **103**, 10424 (1995).
67. F. Bonneté, S. Finet, and A. Tardieu, Second virial coefficient: Variations with lysozyem crystallization conditions, *J. Cryt. Growth* **196**, 403 (1999).
68. J. Wu, D. Bratko, and J. M. Prausnitz, Interaction between like-charged colloidal spheres in electrolyte solutions, *Proc. Natl. Acad. Sci. USA* **95**, 15169 (1998).
69. M. Casselyn, J. Perez, A. Tardieu, P. Vachette, J. Witz, and H. Delacroix, Spherical plant viruses: Interactions in solution, phase diagrams and crystallization of brome mosaic virus, *Acta Crystallogr D Biol Crystallogr* **57**, 1799 (2001).
70. D. N. Petsev, B. R. Thomas, S.-T. Yau, D. Tsekova, C. Nanev, W. W. Wilson, and P. G. Vekilov, Temperature-independent solubility and interactions between apoferritin monomers and dimers in solution, *J. Cry. Growth* **232**, 21 (2001).
71. D. Leckband and J. Israelachvili, Intermolecular forces in biology, *Quart. Rev. Biophysics* **34**, 105 (2001).
72. S. K. Pal and A. H. Zewail, Dynamics of water in biological recognition, *Chem. Rev.* **104**, 2099 (2004).
73. M. Manciu and E. Ruckenstein, Long range interactions between apoferritin molecules, *Langmuir* **18**, 8910 (2002).

74. V. Paunov, E. Kaler, S. Sandler, and D. Petsev, A model for hydration interactions between apoferritin molecules in solution, *J. Colloid Interface Sci.* **240**, 640 (2001).
75. M. Doxastakis, Y.-L. Chen, and J. J. de Pablo, Potential of mean force between two nanometer-scale particles in a polymer solution, *J. Chem. Phys.* **123**, 034901 (2005).
76. S. Brandon, P. Katsonis, and P. G. Vekilov, Multiple extrema in the intermolecular potential and the phase diagram of protein solutions, *Phys. Rev. E.* **73**, 061917 (2006).
77. P. Katsonis, S. Brandon, and P. G. Vekilov, Corresponding-states laws for protein solutions, *J. Phys. Chem.* **110**, 17638 (2006).
78. D. N. Petsev and P. G. Vekilov, Evidence for Non-DLVO hydration interactions in solutions of the protein apoferritin, *Phys. Rev. Lett.* **84**, 1339 (2000).
79. G. S. Adair, A theory of partial osmotic pressures and membrane equilibria with special reference to the application of Dalton's law to hemoglobin solutions in the presence of salts, *Proc. Roy. Soc. London A* **120**, 573 (1928).
80. M. S. Prouty, A. N. Schechter, and V. A. Parsegian, Chemical potential measurements of deoxyhemoglobin S polymerization, *J. Mol. Biol.* **184**, 517 (1985).
81. R. C. Williams, Concerted formation of the gel of hemoglobin S, *Proc Natl Acad Sci USA* **70**, 1506 (1973).
82. D. Elbaum, R. L. Nagel, and T. T. Herskovitz, Aggregation of deoxyhemoglobin S at low concentrations, *J. Biol. Chem.* **251**, 7657 (1976).
83. Z. Kam and J. Hofrichter, Quasi-elastic light scattering from solutions and gels of hemoglobin S, *Biophys. J.* **50**, 1015 (1986).
84. A. P. Minton, Non-ideality and the thermodynamics of sickle cell hemoglobin gelation, *J. Mol. Biol.* **110**, 89 (1977).
85. P. D. Ross, R. W. Briehl, and A. P. Minton, Temperature dependence of nonideality in concentrated solutions of hemoglobin, *Biopolymers* **17**, 2285 (1978).
86. P. G. Vekilov, A. R. Feeling-Taylor, D. N. Petsev, O. Galkin, R. L. Nagel, and R. E. Hirsch, Intermolecular interactions, nucleation and thermodynamics of crystallization of hemoglobin C, *Biophys. J.* **83**, 1147 (2002).
87. E. Antonioni and M. Brunori, *Hemoglobin and Myoglobin in their Reactions with Ligands.* North Holland, Amsterdam, 1971.
88. M. L. Broide, C. R. Berland, J. Pande, O. O. Ogun, and G. B. Benedek, Binary liquid phase separation of lens proteins solutions, *Proc. Natl. Acad. Sci. USA* **88**, 5660 (1991).
89. C. Liu, N. Asherie, A. Lomakin, J. Pande, O. Ogun, and G. B. Benedek, Phase separation in aqueous solutions of lens gamma-crystallins: Special role of gammas, *Proc Natl Acad Sci USA* **93**, 377 (1996).
90. M. Muschol and F. Rosenberger, Liquid-liquid phase separation in supersaturated lysozyme solutions and associated precipitate formation/crystallization, *J. Chem. Phys.* **107**, 1953 (1997).
91. D. N. Petsev, X. Wu, O. Galkin, and P. G. Vekilov, Thermodynamic functions of concentrated protein solutions from phase equilibria, *J. Phys. Chem. B* **107**, 3921 (2003).
92. Q. Chen, P. G. Vekilov, R. L. Nagel, and R. E. Hirsch, Liquid-liquid separation in hemoglobins: Distinct aggregation mechanisms of the $\beta6$ mutants., *Biophys. J.* **86**, 1702 (2004).
93. P. E. Bonnett, K. J. Carpenter, S. Dawson, and R. J. Davey, Solution crystallization via a submerged liquid-liquid phase boundary: Oiling out, *Chem. Commun.*, 698 (2003).
94. O. Galkin and P. G. Vekilov, Mechanisms of homogeneous nucleation of polymers of sickle cell anemia hemoglobin in deoxy state, *J. Mol. Biol.* **336**, 43–59 (2004).
95. M. Shah, O. Galkin, and P. G. Vekilov, Smooth transition from metastability to instability in phase separating protein solutions, *J. Chem. Phys.* **121**, 7505 (2004).
96. N. Choudhury and B. M. Pettitt, Dynamics of water trapped between hydrophobic solutes, *J. Phys. Chem. B* **109**, 6422 (2005).
97. O. Annunziata, N. Asherie, A. Lomakin, J. Pande, O. Ogun, and G. B. Benedek, Effect of polyethylene glycol on the liquid–liquid phase transition in aqueous protein solutions, *Proc. Natl. Acad. Sci. USA* **99**, 14165–14170 (2002).

98. M. L. Broide, T. M. Tominc, and M. D. Saxowsky, Using phase transitions to investigate the effects of salts on protein interactions, *Phys. Rev. E* **53**, 6325 (1996).
99. R. Peters, Noise on photon correlation functions and its effect on data reduction algorithms, in *Dynamic Light Scattering. The Method and Some Applications* (W. Brown, ed.), p. 149. Clarendon Press, Oxford, 1993.
100. M. K. Ceretta and K. A. Berglund, Raman spectroscopy of KH_2PO_4 and other aqueous salt solutions, *J. Cryst. Growth* **84**, 577 (1987).
101. O. Gliko, W. Pan, P. Katsonis, N. Neumaier, O. Galkin, S. Weinkauf, and P. G. Vekilov, Metastable liquid clusters in supersaturated and undersaturated protein solutions., *J. Phys. Chem. B* **111**, 3106 (2007).
102. E. E. Uzgiris and D. C. Golibersuch, Excess scattered-light intensity fluctuations from hemoglobin, *Phys. Rev. Lett.* **32**, 37 LP (1974).
103. C. A. Angell, Formation of glasses from liquids and biopolymers, *Science* **267**, 1924 (1995).
104. H. Tanaka, General view of a liquid-liquid phase transition, *Phys. Rev. E* **62**, 6968 (2000).
105. A. J. Malkin, Y. G. Kuznetsov, and A. McPherson, Defect structure of macromolecular crystals, *J. Struct. Biol* **117**, 124 (1996).
106. S.-T. Yau, B. R. Thomas, O. Galkin, O. Gliko, and P. G. Vekilov, Molecular mechanisms of microheterogeneity-induced defect formation in ferritin crystallization, *Proteins: Struct. Funct. Genet.* **43**, 343 (2001).
107. P. G. Vekilov and F. Rosenberger, Dependence of lysozyme growth kinetics on step sources and impurities, *J. Crystal Growth* **158**, 540 (1996).
108. A. McPherson, A. J. Malkin, and Y. G. Kuznetsov, Atomic force microscopy in the study of macromolecular crystal growth, *Annu. Rev. Biomol. Struct.* **20**, 361 (2000).
109. A. McPherson, A. J. Malkin, Y. G. Kuznetsov, and M. Plomp, Atomic force microscopy applications in macromolecular crystallography, *Acta Crystallogr D Biol Crystallogr* **57**, 1053 (2001).
110. H. Lin, S.-T. Yau, and P. G. Vekilov, Dissipating step bunches during crystallization under transport control, *Phys. Rev. E* **67**, 0031606 (2003).
111. P. G. Vekilov, A. Feeling-Taylor, and R. E. Hirsch, Nucleation and growth of crystals of hemoglobins: Case of HbC, in *Methods in Hemoglobin·Disorders in, Series in Molecular Medicine* (R. L. Nagel, ed.). Humana Press, New Jersey, 2003.
112. A. A. Chernov, L. N. Rashkovich, I. L. Smol'sky, Y. G. Kuznetsov, A. A. Mkrtchan, and A. I. Malkin, Processes of growth of crystals from aqueous solutions, in *Growth of Crystals* (E. E. Givargizov, ed.), Vol. 15. Consultant Bureau, New York, 1986.
113. W. K. Burton, Cabrera, N. and Frank, F. C., The growth of crystals and equilibrium structure of their surfaces., *Phil. Trans. Roy. Soc. London Ser. A* **243**, 299 (1951).
114. O. Galkin, W. Pan, L. Filobelo, R. E. Hirsch, R. L. Nagel, and P. G. Vekilov, Two-step mechanism of homogeneous nucleation of sickle cell hemoglobin polymers, *Biophys. J.* **92**, 902 (2007).
115. M. D. Serrano, O. Galkin, S.-T. Yau, B. R. Thomas, R. L. Nagel, R. E. Hirsch, and P. G. Vekilov, Are protein crystallization mechanisms relevant to understanding and control of polymerization of deoxyhemoglobin S?, *J. Cryst. Growth* **232**, 368 (2001).
116. K. Chen, S. K. Ballas, R. R. Hantgan, and D. B. Kim-Shapiro, Aggregation of normal and sickle hemoglobin in high concentration phosphate buffer, *Biophys. J.* **84**, 4113 (2004).
117. Y. Liu, W.-R. Chen, and S.-H. Chen, Cluster formation in two-Yukawa fluids, *J. Chem. Phys.* **122**, 044507 (2005).
118. S. Mossa, F. Sciortino, P. Tartaglia, and E. Zaccarelli, Ground-State Clusters for Short-Range Attractive and Long-Range Repulsive Potentials, *Langmuir* **20**, 10756 (2004).
119. P. J. Steinhardt, D. R. Nelson, and M. Ronchetti, Bond-orientational order in liquids and glasses, *Phys. Rev. B* **28**, 784–805 (1983).
120. S. Auer and D. Frenkel, Numerical prediction of absolute crystallization rates in hard-sphere colloids, *J. Chem. Phys.* **120**, 3015 (2004).

Application of Discrete Molecular Dynamics to Protein Folding and Aggregation

S.V. Buldyrev

Abstract With the rapid increase in computational speed and memory, simulations of proteins and other biological polymers begin to gain predictive power. However, in order to simulate a folding trajectory of a moderate size protein or an aggregation process of a large number of peptides, traditional molecular dynamics methods based on explicit solvent and accurate force field models still must gain several orders of magnitude in speed. Under these circumstances, simplified models which capture the essential features of the system under study may shed light on the problem in question. One of these simplified methods is discrete molecular dynamics (DMD). DMD replaces the interaction potentials between atoms and covalent bonds by discontinuous step functions. This simplification as well as coarse graining of the model (replacing groups of atoms by one effective bead) and replacing the effect of solvent by varying the strength of inter-bead interactions can speed up simulations sufficiently to generate many folding–unfolding events and to track the aggregation of many peptides. This increase in speed is gained mainly due to the ballistic motion of either secondary structures of the protein or individual peptides. This ballistic motion is a characteristic feature of the DMD method. This chapter will review successes and failures of the DMD method in protein folding and aggregation.

1 Introduction

Protein folding and protein aggregation are very important problems in biology and medicine. In spite of enormous advances in experimental studies of proteins, the problem of identification of the protein native state given its amino acid sequence and the inverse problem of designing a protein with a given native state remain unsolved. Many neurological diseases including prion diseases (such as the notorious Mad Cow disease) and Alzheimer's disease, as well as various genetic disorders are related to protein missfolding and subsequent polypeptide aggregation into

S.V. Buldyrev
Department of Physics, Yeshiva University, 500 West 185th Street, New York, NY 10033 USA,
buldyrev@yu.edu

Buldyrev, S.V.: *Application of Discrete Molecular Dynamics to Protein Folding and Aggregation.*
Lect. Notes Phys. **752**, 97–131 (2008)
DOI 10.1007/978-3-540-78765-5_5

insoluble fibrils [1, 2, 3]. Understanding of these processes is extremely important for prevention and treatment of these diseases. Can molecular dynamic simulations be of use in this area? All-atom molecular dynamic simulations with accurate force-fields and explicit solvent are still too slow to simulate complete folding and aggregation trajectories. Therefore, simplified coarse-grained models, which replace solvent by effective attraction or repulsion of the residues, are needed.

One such approach is discrete molecular dynamics (DMD), which replaces atoms or groups of atoms by hard spheres interacting by discontinuous stepwise potentials. DMD has been proven useful for studies of simple liquids [4, 5, 6, 7, 8, 9, 10, 11, 12], polymers [13, 14, 15, 16, 17, 18, 19], colloids [20, 21, 22], lipid membranes [23], and DNA-histone binding [24]. For a recent mini-review of the DMD applications for protein folding and aggregation, see [25, 26]. Due to its simplicity, DMD is also an ideal aid in teaching thermodynamics, physical chemistry, and polymer physics [27, 28]. Here we review recent works which use DMD in the studies of protein folding and aggregation.

2 Discrete Molecular Dynamics

DMD, also known as discontinuous molecular dynamics or event-driven molecular dynamics, was introduced in 1959 by Alder and Wainwright [4] for simulations of hard spheres. Later it was used by Rapaport [13, 14, 29] for simulation of polymer chains, and finally was adopted for simulations of protein-like polymers [15].

Traditional molecular dynamics [30, 31, 32] approximately integrates Newton's equations of motion of particles interacting via continuous pair potentials (e.g., Lennard-Jones or Coulomb) by updating particles coordinates and velocities at fixed time steps of the order of a few femtoseconds. DMD [31] approximates these potentials by a discontinuous step-functions of interparticle distance r. Thus in DMD, particles move along straight lines with constant velocities until a moment of collision, i.e. a moment of time at which r becomes equal to the point of a discontinuity of the potential (Fig. 1). This discontinuity may be of an infinite height (hard-sphere, or an unbreakable chemical bond) or of a finite size (square well or shoulder). The exact time of the next collision can be obtained by finding a minimal positive solution of the correspondent quadratic equations for all pairs of particles (See Appendix A for details). Next, the velocities of the pair of colliding particles are updated using laws of energy, momentum, and angular momentum conservation. These one scalar and two vector equations are sufficient to find the six unknown components of the velocities after the collision and can be solved exactly by reduction to a single quadratic equation of energy conservation. If this equation has no roots, it means that particles do not have enough kinetic energy to jump out of the square well and they recoil back, as in hard-core collision, without change in kinetic energy. Thus, in contrast to the traditional molecular dynamics, DMD provides an exact solution of the system interacting via given discontinuous potentials with strict (subject only to rounding-off errors) conservation of energy and momentum.

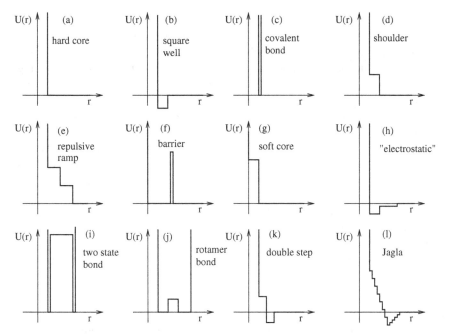

Fig. 1 A collection of DMD potentials used in various studies. (**a**) Hard spheres introduced by Alder and Wainwright [4]; (**b**) square wells and (**c**) covalent bonds were used to study polymer collapse by Rapaport [14]; (**d**) repulsive shoulder used to model hydrophilic interactions or non-native contacts [16, 51]; (**e**) repulsive ramp with two steps used in [87] for auxiliary interactions in hydrogen bond algorithm and with multiple steps in [9, 10, 11] to model liquids with negative thermal expansion coefficient; (**f**) potential barrier used in [21] for ghost particles which occupies no volume but serve as a heat bath; (**g**) soft core is used in all our studies to create an initial configuration of nonoverlapping hard spheres by running at low temperature; (**h**) long-range potential used to model electrostatic interactions in Refs. [61, 99, 106]; (**i**) two-state bond is used to create auxiliary bonds between the backbone beads if they are also linked by a covalent bond [69, 87]; (**j**) complex bond potential to simulate rotamers [61]; (**k**) double-step potential used to simulate liquid–liquid phase transitions [5, 6, 8]; (**l**) multi-step potential approximating Jagla model [38] for water [12, 19]

The distribution of atom velocities after a few collisions converges to the Maxwell distribution. Thus the chance that a pair of colliding particles will increase its potential energy by the step height ΔU is proportional to the flux through the rim of the square well or shoulder of the particles with the kinetic energy larger than ΔU. This flux is proportional to $\exp(-\Delta U / k_B T)$, where k_B is the Boltzmann constant and T is absolute temperature. On the other hand, the probability of entering the well or descending the shoulder is always one. Hence the DMD is equivalent to the Metropolis Monte-Carlo in which the set of moves is not artificial but is equivalent to the ballistic motion of the particles. Hence DMD is suitable for finding dynamic properties of the system, such as diffusion coefficient, viscosity, and time correlation function.

A variety of different DMD potential types have been used to solve various problems in condensed matter physics (Fig. 1). As the sizes of the steps in the discontinuous approximation of a continuous potential approach zero, the DMD trajectory converges to a trajectory of a traditional molecular dynamics for this continuous potential. Since the coordinates and velocities are updated only at the moments of collisions, DMD is especially efficient for dilute systems interacting with very crude potentials such as a hard core plus a single repulsive shoulder or an attractive well. An example of such a system relevant to protein folding is a model of heteropolymer in a vacuum. Interestingly, such a crude approximation of a potential (after all, Lennard-Jones potential is also a crude approximation) is often sufficient to get essential physics and even chemistry right. The ballistic motion of the particles in the DMD is the main feature which allows to speed up the processes of folding and aggregation. The distant parts of the protein and different peptides approach each other ballistically instead of by slow reptation through the surrounding solvent. This helps to increase the computational speed by orders of magnitude. The disadvantage of this speed-up is that one cannot directly predict folding and aggregation rates.

The step potential for a pair of particles of given types A and B can be encoded by a string:

$$A\ B\ r_0\ r_1\ \epsilon_1\ ...\ r_n\ \epsilon_n,$$

where r_0 is a hard core distance and $r_n > r_{n-1} > ... > r_1 > r_0$, are the distances at which the potential has a discontinuity of step ϵ_i. The values of ϵ_i are positive for repulsive shoulders and are negative for attractive wells. If the last ϵ_n is omitted, it means that the particles are linked by the permanent bond whose length can fluctuate between r_0 and r_n.

In addition, DMD can be efficiently used to model chemical reactions since it is possible to change the type of a particle once another particle approaches it within a certain distance and forms a chemical bond. After the reaction, the members of a bonded pair may interact differently with each other and with other particles. In this way, it is possible to model formation of hydrogen bonds and take into account the maximal valence of a given atom [21, 22]. Moreover, this is an effective way to model many body interactions, since the particle type may depend on the particular configuration of its neighbors. Recently, DMD has been applied for modeling physical gels and strong glass-forming liquids using the maximal valence model [21, 22].

It is also easy to implement in DMD various macroscopic objects so long as they are planes or spheres. This can be used for modeling systems in the confined geometry. It is also possible to implement a gravitational force as well as collisions with ghost particles to model a perfect canonical thermostat [16, 21]. Very recently, DMD has been extended to non-spherical objects [33], (ellipsoids [34, 35]) and patchy surfaces which is potentially a very powerful method for modeling protein crystallization [36].

The discontinuous pair potentials are suitable for accurate modeling of dihedral angles by introduction of auxiliary bonds with a small distance between r_{n-1} and r_n connecting the next to the nearest and third nearest atoms along the chain. However, these bonds result in a lot of small interval collisions, which significantly slow down the computation. An alternative approach for efficient modeling of dihedral angles within the DMD algorithm was proposed in [37].

As one can see, DMD is well suited for studies of simple crude models with the goal of understanding the minimal features of the system needed to reproduce a given phenomenon. The examples of successful application of the DMD are modeling fluids with several critical points [5, 6, 7, 8], water-like thermodynamic anomalies [9, 10, 11, 12], anomalous glass transitions in colloids in which the two different glasses (repulsive and attractive) can exist [20], modeling of physical gels and strong glass-formers with maximal valence model [21, 22]. Very recently, a simple model of a non-polar solvent which exhibits the decrease of hydrophobicity upon cooling has been proposed [19].

Simple geometry of the potentials used in the DMD allows us to understand the basic mechanisms of a phenomenon under study. For example, anomalous expansion of water as well as its hydrophobic effect can be reproduced by a spherically symmetric potential with a repulsive ramp (soft core, Fig. 1b) [38]. The rigid hydrogen-bond tetrahedron of the nearest neighbor water molecules corresponds in this model to a hard core, while the more flexible second shell of neighbors represents the soft core. As the temperature increases, some of the particles from the second shell may enter the first shell jumping onto the repulsive ramp. Thus the average distance between the particles reduces and the liquid may shrink upon heating. Analogous effect explains the increase of hydrophobicity upon heating. Small solute particles like alkanes can no longer find sufficiently large cages between solvent particles which become closer to each other as the temperature increases. This effect is the basis for the cold denaturation of proteins [19].

DMD can also be used for accurate prediction of physical and chemical properties of a system, such as the native state of the protein given its amino acid sequence. But in this case one faces a formidable problem of parameterization of the potentials akin to the same problem in traditional MD. Often the parameters of the DMD potential lack any physical meaning (like auxiliary bonds) and must be introduced only to mimic the geometry of the peptide backbone or hydrogen bonds. Also, since using the explicit solvent immediately eliminates all the advantages of the DMD, one must model the hydrophobic and amphiphylic interactions by the effective attraction or repulsion between amino acids or specific atoms. Since the hydrophobic effect is produced by water molecules which form cages around hydrocarbon groups, this effect strongly depends on temperature and other neighboring groups. Therefore, one needs to introduce different potentials for the same atoms in different groups and possibly the three-body interactions by the reaction scheme discussed above. The parameters of these complex interactions can be obtained by means of statistical analysis of the protein data bases, but this requires huge effort and decades of human-years. While in the last two decades a lot of groups were involved in developing traditional MD (CHARMM [39, 40], GROMACS [41], NAMD [42], AMBER [43], LAMMPS [44]), only a handful of researches work on the development of the DMD. Nevertheless, a substantial progress has been made (e.g., the development of the PRIME model by Hall and co-workers [45]).

The bottle-neck of the DMD algorithm is the effective sorting of the collision times. The computational cost of a naive algorithm which computes all the pair collisions and move all the particles after each collision scales as N^3 where N is the number of particles. Several ways of solving this problem are developed [17, 31]. Our sorting algorithm is described in Appendix A. It allows to reduce computational cost to $N \ln N$.

3 Protein Folding

A polypeptide with a uniquely folded 3d conformation (native state) is called a protein. For a long time, biophysicists were puzzled by the following questions: What makes a polypeptide a functional protein? How does an amino acid sequence define a unique 3d structure of a protein? How can one design a protein with a given native sate? How does a protein find its native state in the vast configurational space? If it would proceed by a random search, a simple estimate predicts that the folding time will be larger than the age of the Universe (Leventhal paradox) [46]. For a recent review of these problems see [47].

What is the minimal set of features of a model that ensures that a heteropolymer folds into a unique native state? This question was addressed in the 1990s with help of lattice models (see [47] and references therein). In a sense, this approach was not unlike the minimalistic studies of Picasso, who created a drawing of a bull with a minimal set of features, which however still allowed a spectator to recognize it [48]. Biologists usually do not appreciate this approach, and so in order to be helpful, physicists must try to move in the opposite direction, from Picasso to Velasquez.

Today it becomes clear [47] that a random heteropolymer does not have a unique native state. Its potential energy landscape has many deep minima, the difference between which are just a few $k_B T$. Thus a random heteropolymer will fold into one of these deep minima. The studies of the lattice models of heteropolymers show that in order for the protein to have a unique native state, its energy must be by several standard deviations lower than the minima of the rest of the basins. It is possible to implement an artificial mutation process based on the Metropolis algorithm which maximizes the Z-value, i.e. the ratio of the difference between the energy of the native state and the energy of a typical basin to the standard deviation of the energy distribution of the basins. Lattice heteropolymers designed in such a way fold into the native state given by a contact map, i.e. the matrix of contacts between the amino acids occupying the adjacent lattice sites in the native state. The success of lattice models suggests that the biologically active proteins are the result of natural selection, which has gradually increased the stability of the primitive pre-biotic proteins [49].

4 The One-Bead Go Model

The next step toward a more realistic picture of a protein would be to design an off-lattice model which folds to a prescribed globular native state. A natural candidate for this model would be a bead-on-a-string model which interacts via square well potentials [13, 14, 29]. The values of the potentials may be taken from an effective matrix of interactions derived from the probabilities of amino acids to be close to each other in the native state of the existing proteins [50]. This model can be very efficiently studied by the DMD. The initial globule can be created by a collapse of a homopolymer, interacting via identical attractive square wells. The initial sequence of the protein is a random sequence of twenty letters and then it is changed by mutations maximizing the energy gap between the energy of the native state and the energies of random contact maps representing missfolded states as it has been done for lattice models. Our studies have shown that this approach does not work. The bead-on-a-string model has too many degrees of freedom and too many contacts per amino acid. The number of contacts reaches ten for the beads in the central core of the bead-on-a-string model while in the lattice model it is only four. So it is impossible to design a protein-like sequence of sixty beads using only the twenty-letter code.

We have found [51] that the Go model [52, 53, 54] which uses 60×60 matrix of interactions (Fig. 2) for a sixty-bead polymer works very well. In the Go model, the beads which are within a certain distance in the native state attract to each other while those that are further away repel. Thus, the native state is by definition the ground state of the model and its energy gap with a randomly missfolded globule is of the order of n_c, where n_c is the number of native contacts. We find that the Go model of a small globular heteropolymer always folds into a native state near the folding temperature, T_f. Moreover, this happens in a reversible way, so that the polymer folds and unfolds many times during the simulation. At $T = T_f$, the folded and unfolded conformations are equally populated and separated by a significant energy gap which is about one half of the total number of native contacts (Fig. 3). This bimodal distribution is a characteristic of the first-order phase transition in which the two phases (liquid and crystal) may coexist and the potential energy gap between the two phases is proportional to the number of molecules in the system. This is in sharp distinction to the behavior of a flexible homo-polymer near the theta point, which undergoes a second-order phase transition and has a unimodal distribution of energies.

Interestingly, the models produced by the Go algorithm from a collapsed state of a homoplymer often have intermediate states in which a tail consisting of a significant number of beads is detached from the rest of the folded globule. The partially folded states comprise an intermediate bump on the potential energy distribution. Cutting away this tail yields a perfect two-state folder, which represents the majority of the small proteins.

As the temperature of the system is reduced below the folding temperature, the energy distribution becomes unimodal, with the probability of being in the folded state dominating (Fig. 4). However, if the temperature of an unfolded state is reduced below a certain value, which is about 70 % of the folding temperature [51], the polymer may never find its native state and may be trapped forever in a missfolded state. This phenomenon is analogous to the glass transition in the supercooled liquids, in which the nucleation of the crystal can be avoided by fast quenching. For protein A, this phenomenon is observed at $T = 0.62$, which is about 80% of $T_f = 0.765$.

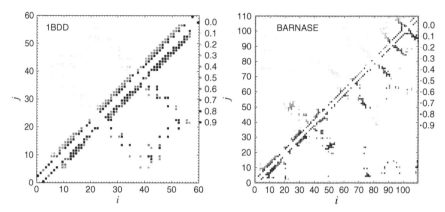

Fig. 2 A Go interaction matrix (contact map) constructed for the B domain of staphylococcal protein A (1BDD) and barnase (1BNR) using their native states taken from the Protein Data Bank. The amino acids for which C_β atoms in the native state are separated by less than 7.5 Å attract to each other via a square well interaction (Fig.1b) with $r_0 = 3.459$ Å , $r_1 = 7.5$ Å , and depth $\epsilon_1 = -\epsilon$, while other amino acids repel from each other via a square shoulder interaction (Fig. 1d) with the same r_0 and r_1 and height $\epsilon_1 = +\epsilon$. The 1BDD protein has 3 α-helices held together by 45 long-range contacts. The total number of native contacts is 160 which corresponds to the ground state potential energy of -160ϵ. A particular structure of the contact map can vary a lot. For example, SH3 domain has 3 β-hairpins. An artificial globule constructed by the homopolymer collapse has neither α-helixes nor β-sheets, but usually have about cN native contacts where N is the number of monomers and c varies between 2 and 3. The number of long-range contacts is about 0.25–0.3 of the total number of contacts. Shades of gray indicate the probabilities of contacts in the folded (*lower triangle*) and unfolded (*upper triangle*) states at $T = T_F = 0.765$. In the unfolded state, all long-range contacts have very low probability, while the secondary structure is already partially formed. In the folded state, many long-range contacts are formed with probability larger than 0.8. These contacts form putative folding nucleus. The energy gap between the folded and unfolded state is 54ϵ. The analogous Go model for barnase has 316 native contacts which is two times larger than for protein A. This Go model also folds cooperatively into the native state but the energy gap between the folded and unfolded states is two times wider than for protein A (107ϵ). So at $T_f \approx 0.8$ barnase can undergo only few transitions between folded and unfolded states in 10^7 time units. The *upper triangle* shows contact probabilities for the unfolded state. The *lower triangle* shows the contact probabilities in the folded state. The protein consists of two α-helices and three β-hairpins. While the α-helices are well formed in the unfolded state the β-hairpins are not present. In order for this protein to fold, the β-hairpins must form cooperatively

This model seems to explain the Leventhal paradox: near the folding temperature, the secondary structure of the unfolded chain is already partially formed with about 50% of the native contacts (mainly short ranged) in place (Fig. 2). The individual elements of the secondary structure are not stable because its potential energy differs from the unstructured conformations only by few $k_B T$. However, at this point the polymer has already lost an immense number of degrees of freedom and acts like a collection of a few secondary structural elements. Once the few critical long distance contacts (the folding nucleus) [55] are formed, the partially folded secondary structural elements come together and the polymer quickly descends into its native state. This process is similar to the formation of the critical nucleus in the first-order

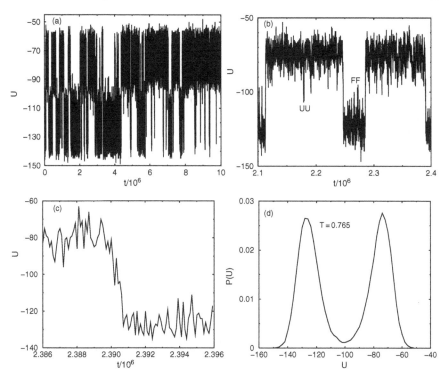

Fig. 3 (**a**) Potential energy versus time for a two-bead Go model of 1BDD protein near $T = T_F = 0.765$. The graphs for the one-bead Go models of two-state proteins look similar. About 60 folding and unfolding events are visible. (**b**) In addition, there are many unsuccessful attempts to fold (UU events) or unfold (FF events). (**c**) Each folding event takes about 600 time units which is about 300 times faster than average time spent in the folded or unfolded state. (**d**) The probability density of the potential energy is bimodal with equally populated folded and unfolded states. The average energy of the unfolded state is -73ϵ which corresponds to the existence of approximately one half of all native contacts

phase transition. During this stage the remaining 50% of the native contacts are formed [55, 56]. However, the discussed folding scenario can be an artifact of the Go model, in which the non-native contacts do not attract to each other and hence the hydrophobic collapse preceding the formation of the secondary structure cannot be observed. Adding small hydrophobic attraction between the non-native contacts to the Go interactions changes the folding scenario [57, 58]. In this case, the protein first undergoes the hydrophobic collapse into a molten globule state, in which the secondary structure is only weakly formed. If the attraction of non-native contacts is significantly weaker than the attraction of native contacts, the molten globule reorganizes itself into the native state in a first-order-like transition, but the folding process is much slower than in the case when the non-native contacts are assigned zero or even positive (repulsive) energy. It is clear that both scenarios can take place in vitro and in vivo depending on the properties of the protein and its environment.

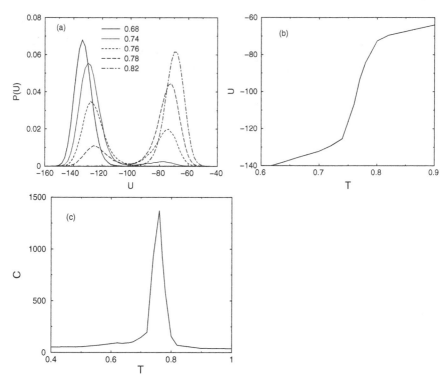

Fig. 4 (**a**) Probability density of the potential energy for different temperatures in the vicinity of the folding transition for the 1BDD protein. The populations of the folded and unfolded states change dramatically while the temperature changes from 0.7 to 0.78. Outside this temperature interval the distribution becomes unimodal. Thus the temperature interval of the folding transition corresponds to about 40 K in the physiological temperature scale. Accordingly, the average potential energy (**b**) dramatically changes with temperature and the heat capacity (**c**) has a sharp maximum near folding transition. All these are features of the first-order phase transition. This behavior is typical for the one-bead and two-bead Go models

The transition time during which the trajectory descends from an unfolded to a folded state is by several orders of magnitude smaller than the average folding time, i.e. the time during which the polymer explores the vast ensemble of the unfolded states. This feature is also observed in the folding of all-atom models with explicit solvent. Actually, it is the basis of the Pande's Folding at Home project [59]. In this approach, hundreds of thousands participants world-wide run the simulations as screen-savers on their personal computers. Average folding time of a small protein is of the order of one hundred microseconds, while a typical simulation time is about 10 ns. In contrast, the descent of the protein from the transition sate in which the critical nucleus is formed into the native state is within the simulations reach. Thus in one case out of ten thousand, a lucky participant may observe an actual folding. Using these results Pande and coworkers can determine the folding rate, an experimentally verifiable quantity.

As we see, DMD helps to explain a lot of features of a real protein folding in simple terms. In summary, the DMD Go model confirms that the protein folding has many features of the first-order phase transition, in particular nucleation [60]. Of course, since the go model explicitly uses the information on the native state, it does not bring us closer to the "Holy Grail" of computational biology, which is an ab initio folding of a protein given only its amino acid sequence. Can pair potentials, even with correct geometry of the peptide backbone and the side chains, but without the explicit solvent, be successful in this quest? The answer to this question is still unknown. The success of folding of the trp-cage miniprotein by the DMD [61] and by other molecular dynamics methods is probably due to a very specific sequence of this protein with several prolines which make the backbone especially rigid. We will review this study below. In reality, amino acids interact not only with each other but also with the surrounding water, which forms cages around the hydrophobic amino acids and makes hydrogen bonds with polar amino acids. Thus, solvent creates effective many-body interactions among amino acids, i.e. changes the amino acid properties by the effects of their neighbors. Thus solvent creates many more than 20 effective amino acid types.

5 Transition States of Realistic Proteins

Can the Go model predict certain experimentally verifiable features of real proteins, such as their transition state ensembles, the folding pathways, and the presence of intermediates? In order to answer this question, we take [62] a well-studied protein with a known native state, Src SH3 domain, and create a Go interaction matrix (Fig. 2), assuming that if C_β atoms of the side chains are less than 7.5 Å apart in the native state, they attract with the square well of diameter 7.5 Å and potential energy $-\epsilon$, while if they are farther apart they repel from each other as hard-cores at the same distance. The later rule insures that the protein cannot form non-native contacts at all. For glycines, which do not have side chains, we use C_α instead of C_β.

Interestingly, the bead on the string model with this Go matrix and all equal bond lengths do not fold into a native state, so we have to specify the bond lengths as the distances between subsequent C_β atoms in the native state. A version of this model, in which the distances between C_α atoms in the native state specify the Go interaction matrix, does not fold into a native state cooperatively. This result indicates that the conformations of the side chains rather than those of the backbone determine the specific nature of the amino acid interactions. The model shows a much more cooperative transition than the previously studied model of an artificially constructed globule [51, 55, 56], although the SH3 domain has similar length, $N = 56$, and number of native contacts, $N_c = 160$. The distribution function of potential energy has a wide gap between the sharp maxima corresponding to the native and unfolded states, separated by a deep minimum. Similar behavior is observed for many other short proteins (Fig. 3).

If one assumes that the potential energy (number of native contacts) is a good reaction coordinate, then the conformation corresponding to the minimum in the energy histogram must belong to the free-energy barrier separating the native and unfolded states and thus constitute the transition state ensemble (TSE), which is by definition a set of conformations that with 50% chance fold within the typical transition time and with 50% chance unfold. In the SH3 domain Go model, the ratio of the transition time to the time for which protein resides in the native and unfolded states is even smaller than in the artificial globule studies. We find that the majority of the conformations near the minimum of the potential energy histogram do not correspond to actual folding or unfolding events but represent the unsuccessful attempts to fold or unfold. We call these attempts folded-folded (FF) and unfolded-unfolded (UU) events. A FF event is a part of the trajectory, which originates below the maximum of the distribution characterizing the folded states, reaches the minimum of the distribution and without ever reaching the maximum characterizing the unfolded states descends below the maximum characterizing the folded states. The UU events are defined analogously (Fig. 3b). We hypothesize that both FF and UU events have structural similarities to the TSE. We also hypothesize that the difference between the UU and FF events is that in FF events the folding nucleus is not destroyed, while in the UU events the folding nucleus does not form. Thus the contacts which represent the folding nucleus must be those with the maximum positive difference of their probabilities to be in FF and UU events. (Note that the average number of contacts in FF events is smaller than in the UU events, because we sample the UU events below the minimum of the distribution, while the FF events are sampled above the minimum of the distribution.)

We find [62] that some contacts are significantly more abundant in FF events than UU events. In the Go model of the Src SH3 domain, there are about 20 such contacts and all of them belong to the two distinct clusters of long-distance contacts in the contact map: one characterizes the contacts between the termini of the protein while another characterizes the contacts between the distal hairpin and the RT-loop. However, the absolute values of the probabilities to find contacts between the termini are not very high in the FF event, so we conclude that these contacts do not characterize the TSE and thus do not belong to the folding nucleus. On the other hand, the contacts between the RT-loop and the distal hairpin occur in more than half of FF events, so we assume that they do belong to TSE and thus form the true folding nucleus.

We also produce the P-fold analysis as an additional test of the putative folding nucleus. The P-fold analysis consists of randomly changing velocities of the amino acids in a certain conformation and then performing a simulation for a time interval which is significantly longer than a typical transition time but is significantly shorter than the time of staying in the folded or unfolded states. Then we count the fraction P of trials in which the protein ends up in the folded state. The P-fold analysis confirms our identification of the putative TSE and folding nucleus. The amino acids belonging to this putative folding nucleus form contacts between the distal hairpin and the RT loop. This conclusion is however not confirmed in the later studies in which the putative TSE is identified by the all-atom importance sampling MD [63].

These contacts are not present in any of the 51,000 conformations sampled in the vicinity of the free-energy barrier on the two-dimensional map which uses radius of gyration and potential energy as reaction coordinates. On the other hand, the conformations with termini contacts are abundant in this putative TSE. This discrepancy questions both the ability of simple Go models to make reasonable predictions of the true folding kinetics and the validity of the all-atom modeling based on importance sampling molecular dynamics (in these studies no actual folding trajectories are obtained). Only experiment can bring the final verdict.

This discrepancy is consistent with the idea of multiple folding pathways which we found for the SH3 domain at low temperatures and some other short proteins using the one-bead Go model [62, 64]. At high temperatures, within 20% range from T_f, the protein can fold via two pathways either forming contacts between distal hairpin and RT loop or (with smaller probability) forming contacts between C-N termini. At high temperature, both pathways are fast and the folding is optimal at about $0.85T_f$, which is consistent with experimental observations. However, at $T < 0.60T_F$, the protein can be trapped in the intermediate state with the contacts between the termini formed prematurely. In order to proceed to the folded state, the protein must first break these contacts which requires certain activation energy. That is why, the time spent in the trap diverges at low T following the Arrhenius law.

In addition, we study unfolding of nine other proteins which are known to have folding intermediates. In general, our results agree with the experimental findings: namely for the two-state folding protein like SH3 and Im9 domains, the Go model predict cooperative folding with no intermediates, while for the proteins with intermediates, including Im7 which is homologous to Im9 but is known to have intermediates, the Go model also predicts intermediates. This remarkable success of Go models suggests that it is the topology of the native state rather than the amino acid sequence that determines the kinetics and thermodynamics of the globular proteins.

A different variant of the one-bead model with Go interactions has also been used for DMD simulations of various folding pathways in α-helical [57, 58] and β-stranded proteins [65]. These simulations use a pseudo dihedral angle potential which creates a chirality bias toward right-handed α-helices [16]. In these models all the contacts (both native and non-native) can form, but the interaction energy of the native contacts is larger than that of the non-native ones. By varying this energy gap between native and non-native contacts, various folding scenarios are observed. For the large energy gap, the protein can not only quickly fold into a native state, but can also be trapped in the partially missfolded conformations. For the low energy gap, the protein first collapses into a compact disordered structure similar to a molten globule [66] and then slowly folds into a native state. Thus, these studies show that both scenarios are possible for protein folding. In the first scenario, parts of secondary structural elements form in the unfolded state [67], which then fold into the native state. During this process missfolded conformations can form which must unfold for the successful folding into the native state. In the second scenario, proteins collapse into a disordered globule, which later fold into the native state either in a cooperative transition or via non-obligatory intermediates.

However, a third scenario in which the hydrophobic collapse happens simultaneously and cooperatively with the formation of the secondary structure is probably the most likely one. Modern theory of hydrophobic interactions [68] predicts complete dewetting of large polymer globules and formation of the water phase boundary around globules exceeding 1 nm in diameter. The formation of such a globule is a two-state, first-order-like phase transition even for a flexible homopolymer. Recently, this transition has been demonstrated by the DMD simulations of a hard sphere polymer in a water-like Jagla solvent [19]. The formation of the relatively weak hydrogen bonds between the backbone carbonyl and amide, are crucial for formation of the secondary structure, is unlikely if the backbone is surrounded by water molecules which can form much stronger hydrogen bonds with the backbone groups. Expulsion of water by the hydrophobic dewetting caused by the hydrophobic side chains enhances the formation of the backbone hydrogen bonds which in its turn serves as positive feedback for the collapse. Unfortunately, this scenario is impossible to simulate replacing the effect of solvent by effective pair potentials between protein atoms.

6 The Two-Bead Go Model

We see that the one-bead Go model is too flexible and needs adjustments of the distances between C_β atoms in order to take into account the geometry of the backbone and the side chains. Thus, we take a next step "toward Velasquez" and explicitly add the side chains (each represented by a single bead C_β) to the backbone, which is still represented by the chain of C_α [69]. In order to reduce the flexibility of the backbone, we add auxiliary bond linking next to the nearest C_α atoms along the backbone. These bonds make rigid isosceles triangles $C_{\alpha,i-1}C_{\alpha,i}C_{\alpha,i+1}$ with the angle at $C_{\alpha,i}$ of approximately $96°$ and the distance between $C_{\alpha,i-1}$ and $C_{\alpha,i}$ of approximately 3.8Å. The $C_{\beta,i}$ is attached to the $C_{\alpha,i}$ bead by a covalent bond of 1.53Å and by two auxiliary bonds linking it with $C_{\alpha,i-1}$ and $C_{\alpha,i+1}$ in order to keep the covalent bond approximately orthogonal to the plane of the triangle $C_{\alpha,i-1}C_{\alpha,i}C_{\alpha,i+1}$. Thus, the entire protein consists of imperfect tetrahedra connected by their edges in such a way that $C_{\beta,i}$ and $C_{\beta,i+1}$ point in opposite directions. The methods for determining the Go interaction matrix is the same as in the one-bead model (Fig. 2). Note that glycines lack C_β and we compute their native contact map using their C_α coordinates. Both covalent and auxiliary bonds can fluctuate by a few percent within a square well. This flexibility intends to mimic the actual statistics of the backbone geometry. Thus the model can form approximate β sheet conformations as well as α-helixes.

We use the two-bead model to study the SH3 domain TSE [69] performing P-fold analysis [70] and FF–UU event analysis as in [62]. As in the one-bead model, the SH3 domain folds in a highly cooperative two-state transition. The transition state appears to be the same in both one-bead and two-bead models. In addition, we simulate Φ values using a virtual screening method which we develop for the

Go model and compare them with the experimental Φ values. The virtual screening method computes the shifts in the Gibbs potentials ΔG_U, ΔG_T, and ΔG_F, of the unfolded, transitional, and folded states of the protein, produced by a "mutation" of an amino acid which turns off the interaction energies of this amino acid with its neighbors in the native state. It is assumed that such point mutations usually do not alternate the topology of the native and transition states. The change in the Gibbs potential of a certain state X due to mutation is defined as $\Delta G_X = -k_B T \ln\langle\exp(-\Delta E_X/k_B T)\rangle$, where ΔE_X is the change in potential energy due to the mutation, and $\langle...\rangle$ determines the average over all the observed conformations in the considered state. Finally, we compute the Φ value of the amino acid as $\Phi = (\Delta G_T - \Delta G_U)/(\Delta G_F - \Delta G_U)$, which compares the increase in the height of the free energy barrier to the decrease in the protein stability [71].

Note that no additional simulations are made in the virtual screening method but we use the conformations obtained in the original Go-model. Thus, the Φ values are fully determined by the contact map distributions of the transition state. Amino acids which have the same number of contacts in the transition state as in the native state have $\Phi \approx 1$. The amino acids which do not form native contacts in the transition state have $\Phi \approx 0$. Note that in the Go model the Φ values can be only within the range $[0, 1]$ while in experiments they can be sometimes larger than one or even negative; thus, one cannot expect high correlation between the simulated and the experimental values. Indeed, we found the correlation coefficient $r = 0.58$. Particularly disturbing is the high probability of the contact between amino acids L24 and G54 from the RT loop and distal hairpin in our simulations and their low experimental Φ values. On the other hand, these are the most conserved amino acids in the SH3 family, so it seems that they should be important for the protein stability. In fact, the mutation of G54 in experiments destabilizes the native state but reduces the transition state barrier, thus it has a negative experimental Φ which may indicate its participation in the transition state. This may be explained by the presence of the backbone hydrogen bond between G54 and L24, which is not destroyed in mutations.

To further test the role of this particular contact in the Go model, we replace it with a permanent bond. The crosslinking of these amino acids significantly increases the population of the transitional state, while the crosslinking of the termini does not change it significantly. Nevertheless, the question of the relative importance of the two folding pathways in SH3 domain remains open. A recent article of Lam et al. [72] shows that by changing the relative strength of the Go-interactions in the different segments of the two-bead model of the SH3 domain, one can dramatically increase the probability of the folding pathway via formation of the contacts between the termini without the change in the stability of the native state and cooperativity of the folding transition.

The secondary structure of the SH3 domain consists of only β-hairpins and does not have α-helices. As an example, we test the two-bead model by folding two small proteins, the B domain of staphylococcal protein A (PDB access code 1BDD) which has three α-helices and barnase (PDB access code 1BNR) which has both α-helices and β-hairpins (Fig. 2). It appears that 1BDD folds cooperatively in a two-state transition the same way as the SH3 domain (Figs. 3 and 4). At the folding temperature, $T_f = 0.765$, both folded and unfolded states are equally populated and the protein undergoes rapid folding and unfolding transition. In the unfolded state, the α-helices are already well formed but the long-range contacts are not present. Thus during folding, the α-helices come together and form the tertiary structure. The folding nucleus belongs to the set of long-range contacts.

Similar situation is observed in barnase, which cooperatively folds at $T_f \approx 0.8$ in a two-state process. This result coincides with the conclusions of [64] in which the absence of the intermediates for barnase has been reported based on a few unfolding trajectories of the one-bead Go model. We study the structure of folded and unfolded states at $T = 0.8$. In the unfolded states, the α-helices are well formed but some of the β-hairpins are not present. This is clear because the α-helices are formed by the short-range contacts and thus their formation costs much smaller entropy loss than the formation of β-hairpins for which the contacts have longer range. Accordingly, the protein folds simultaneously with the two central β-hairpins which are likely to be a part of the TSE. There is an experimental evidence that barnase has an on-pathway folding intermediate [73]. The fact that this intermediate aggregates may indicate that it has exposed β-strands. This is consistent with our simulation results.

However, in reality, the formation of the β-hairpins may precede the α-helices. The two-bead Go model does not take into account the entropy of the side chains. Accordingly, β-hairpins may have larger entropy than α-helices and may form at higher temperatures. By changing the energy [72] or the range of the Go-interactions, one can change the folding pathway and reverse the order of formation of α-helices and β-hairpins.

The two-bead Go model is now publicly available for the P-fold analysis of the arbitrary protein structures [70]. It can be used for studies of large-scale conformational dynamics of long proteins consisting of thousands of residues and their binding [74]. It has been employed in the study of the conformations of the denatured proteins [67]. The nature of the Go interactions leads to the abundance of the native secondary structural elements in the denatured states. However, this result may depend on the relative strength of the native and non-native contacts as discussed in [57, 58]. Very recently the two-bead Go model has been used for simulation of histones binding to DNA [24]. Each DNA nucleotide has been modeled by three effective beads.

7 The Two-Bead Model with Hydrogen Bonds: Studies of Protein Aggregation

It is well known that many genetic neurological diseases are caused by aggregation of proteins into insoluble fibrils formed by the β-sheets crosslinked by hydrogen bonds [1, 2, 3]. While an isolated protein can still fold into its native state, the proteins in concentrated solutions can attach to each other by the exposed β-strands, and then find a new deeper free-energy minimum corresponding to the insoluble fibrils with a regular structure. Thus, the fibril formation in many aspects is similar to crystallization except that in protein crystallization the proteins remain in the native state and are attached together by relatively weak side chain interactions. Understanding of the mechanism of this phenomena is of crucial importance in developing drugs which would prevent protein aggregation and thus stop the development of such devastating neurological syndromes as Alzheimer's disease, Parkinson's disease, Huntington's disease, and prion diseases [1, 2, 3, 75, 76].

As a first step in modeling aggregation, we introduce hydrogen bonds into our two-bead model in addition to the Go interactions [77, 78]. The amino acid in each peptide is identified by its index i, which is its number starting from the N terminus. We assume that if a certain pair of amino acids i and j within the same peptide interacts via an interaction potential, determined by the native state, it interacts with the same potential even if its members i and j belong to different peptides. Also, we assume that the hydrogen bonds can form between the amino acids in the same peptide and between the different peptides, except that the hydrogen bonds between amino acids i and j from the same peptide are forbidden if $|i - j| < 3$. This rule is derived from the extensive studies of the structures in the Protein Data Bank (PDB).

In reality, the backbone hydrogen bonds are formed between nitrogens and oxygens from the carbonyl groups of the backbone. Thus each amino acid can form at most two backbone hydrogen bonds one by donating a hydrogen by the nitrogen and another by accepting a hydrogen by the carbonyl. The geometry of the peptide backbone is such that these two bonds must be approximately parallel. In the two-bead model we do not have carbonyls and nitrogens so we introduce the effective bonds between C_α beads. Each C_α is allowed to have two hydrogen bonds. So, each C_α keeps track of the number of hydrogen bonds and the amino acids linked by these bonds. If two beads $C_{\alpha,1}$ and $C_{\alpha,2}$ have no hydrogen bonds then they will always form a new hydrogen bond as soon as they come to a distance of 5 Å. If $C_{\alpha,2}$ already has a hydrogen bond with $C_{\alpha,1}$ and comes within 5 Å of the bead $C_{\alpha,3}$, the hydrogen bond between $C_{\alpha,2}$ and $C_{\alpha,3}$ can be formed only if the distance between $C_{\alpha,1}$ and $C_{\alpha,3}$ is between 8.7 Å and 10 Å. In addition to this hydrogen bond, an auxiliary bond is formed between $C_{\alpha,1}$ and $C_{\alpha,3}$ which is modeled by the infinite square well of this width (between 8.7 Å and 10 Å). This auxiliary bond keeps the angle between the two hydrogen bonds close to 180° during the time of the existence of these hydrogen bonds. If the beads linked by a hydrogen bond approach 5 Å distance, the hydrogen bond breaks according to the normal rules of the DMD and the auxiliary bonds which may exist between them and their partners break simultaneously without any energy loss. We take the energy of the hydrogen bond $\epsilon_{HB} = 3\epsilon_{Go}$, because the hydrogen bond interactions are much stronger than the hydrophobic interactions between the side chains. Of course, the role of water, which may also form the hydrogen bonds with the backbone, is totally neglected. Surprisingly, this highly simplified model, while keeping the native structure of a single protein, produces perfect β-sheets between different peptides.

We decided to simulate the aggregation of the SH3 domain [77] because we had already studied its folding and because aggregation does not depend on the amino acid–specific interactions and thus can in principle involve any protein. We placed eight identical proteins in the simulation box and raised the temperature, to completely unfold them. Then we ran long equilibrium simulations at different temperatures. The proteins moved ballistically around the box, so the chance that they can soon come in the vicinity of each-other was very high. At first they formed disordered oligomers which later transformed into β-sheet fibrils. The fastest fibril formation happened near folding temperature T_F, when the individual proteins frequently unfolded but retained significant amount of the secondary structural elements in the unfolded state. Accordingly, the SH3 domain proteins formed parallel and antiparallel β-sheets between the RT loops which produce a sort of a barrel with an axis parallel to the direction of the hydrogen bonds. The terminal regions of these proteins remained highly disorganized. The fact that this model can correctly predict the domain swapping between two proteins indicate that even this very simple model can correctly capture some features of protein aggregation [77, 78].

Similar results have been achieved in the simulations of β-amyloid peptides whose native state has been modeled by the α-helices [79]. Aggregation of the amyloid-β peptide is believed to be the leading cause of the Alzheimer disease. Interestingly, the peptides first form unstructured oligomers with high degree of α-helical structure still present, and only later do they organize themselves in the perfect multi-layered β-sheets. This finding is in accordance with the present hypothesis that the death of neurons in the Alzheimer disease is caused not by the ordered amyloid fibrils accumulated in the plaques, but by short-lived intermediate oligomers which are precursors for fibrillization [80, 81]. These oligomers are highly mobile and can attach themselves to the cell membranes and probably puncture them. Similar β-fibrils have been obtained by Hall and coworkers in the aggregation of polyalanine [82, 83] and polyglutamine [84] by an intermediate resolution protein model similar to the four-bead model discussed next. The polyglutamine fibril formation is the molecular basis of Huntington disease.

8 The Four-Bead Model: Studies of the α-Helix-to-β-Hairpin Transition

The next step toward a realistic protein model that can fold into a native state without explicitly specifying its topology by the Go interaction matrix is to create a model which can reproduce spontaneous formation of the secondary structure, i.e. α-helices and β-sheets. The role of secondary structure formation is crucial in understanding the protein folding and aggregation. In α-helices, all the backbone hydrogen bonds are used, thus they cannot aggregate into fibrils, while in β-sheets only half of the hydrogen bonds are used and thus they can easily aggregate. In order for the peptide with a significant amount of α-helices to aggregate, the α-helices must spontaneously transform into β-sheets [85].

Go models predict that about 50% of the secondary structure is formed in the unfolded sate before the cooperative folding takes place. The secondary structure in the unfolded state is not stable because in the presence of water, backbone can form hydrogen bonds with surrounding water molecules which are stronger than the interpeptide hydrogen bonds. The hydrogen bond $C=O\cdots H\text{–}O\text{–}H$ between a carbonyl group and a water molecule is about 21 kJ/mol, while the carbonyl–nitrogen hydrogen bond $\text{–}C=O\cdots H\text{–}N\text{–}$ is only 8 kJ/mol. Once the folding nucleus is formed and cooperative folding takes place due to hydrophobic collapse, the water molecules are expelled from the interior of the protein and the protein backbone has no other choice but to form hydrogen bonds between carbonyls and nitrogens. Thus, the secondary structure becomes stable in the folded state. This argument suggests that it is impossible to create an accurate protein model without taking into account the local environment of hydrogen bonds. More advanced models of hydrogen bonds, which can be turned on and off depending on the conformation of the protein in the vicinity of the carbonyl and amide groups, are currently being developed by several research groups.

The four-bead model that we have developed is similar to the PRIME model of Carol Hall and co-workers [86]. This model correctly reproduces the backbone geometry. It replaces each amino acid by a rigid tetrahedron of four beads: N, C_α, CO, and C_β (Fig. 5). In glycines C_β is absent. The rigidity is achieved by three covalent bonds, $N\text{–}C_\alpha = 1.56$ Å, $C_\alpha\text{–}CO = 1.51$ Å, and $C_\alpha\text{–}C_\beta = 1.53$, and three auxiliary bonds, $N\text{–}CO = 2.44$ Å, $C_\beta\text{–}N = 2.44$ Å, and $C_\beta\text{–}CO = 2.49$ Å. The tetrahedra representing amino acids i and $i+1$ are linked together by the rigid planar quadrilateral (plate) $C_{\alpha,i}\text{–}CO_i\text{–}N_{i+1}\text{–}C_{\alpha,i+1}$ formed by the peptide bond $CO_i\text{–}N_{i+1} = 1.33$ Å linking two amino acids together, two covalent bonds $C_{\alpha,i}\text{–}CO_i$, and $N_{i+1}\text{–}C_{\alpha,i+1}$ which participate also in the correspondent tetrahedra, auxiliary bonds $C_{\alpha,i}\text{–}N_{i+1} = 2.41$ Å, $CO_i\text{–}C_{\alpha,i+1} = 2.43$ Å, and the diagonal $C_{\alpha,i}\text{–}C_{\alpha_i+1} = 3.78$ Å, which maintains the quadrilateral rigidity. Note that $C_{\beta,i}$ and $C_{\beta,i+1}$ atoms in the adjacent amino acids point in the opposite directions. The beads N, C_α, CO, and C_β are modeled by hard spheres of radii 1.69, 1.76, 1.75, and 1.54 respectively.

The amino acid tetrahedra can freely rotate (like doors around hinges) around $N\text{–}C_\alpha$ and $C_\alpha\text{–}CO$ bonds forming the two Ramachandran dihedral angles Φ and Ψ,

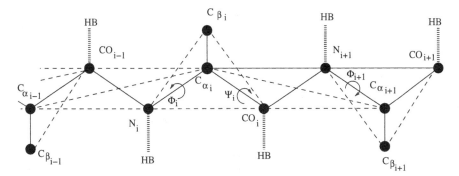

Fig. 5 The four-bead model of the protein backbone. *Bold lines* show covalent bonds, *dashed lines* show auxiliary bonds which helps maintain correct backbone geometry, and *thick broken lines* show possible hydrogen bonds

respectively, between the plane $N_iC_{\alpha,i}CO_i$ and the adjacent quadrilaterals $C_{\alpha,i-1}CO_{i-1}N_iC_{\alpha,i}$ and $C_{\alpha,i}CO_iN_{i+1}C_{\alpha,i+1}$. The sequence of Ramachandran angles $\Psi_1, \Phi_2, \Psi_2, ... \Psi_{n-1}, \Phi_n$ completely determines the backbone conformation. The absolutely planar β-strand conformation corresponds to all $\Psi_i = 0$, $\Phi_i = 0$. Mathematically, the Ramachandran angles are determined by the sequence of the four successive backbone bond vectors, $a_1=CO_{i-1}N_i$, $a_2=N_iC_{\alpha,i}$, $a_3=C_{\alpha,i}CO_i$, and $a_4=CO_iN_{i+1}$:

$$\Phi = \pm\mathrm{acos}\left(\frac{[a_2 \times a_1] \cdot [a_2 \times a_3]}{|a_2 \times a_1||a_2 \times a_3|}\right),$$

where the sign coincides with the sign of $([a_1 \times a_3] \cdot a_2)$ and

$$\Psi = \pm\mathrm{acos}\left(\frac{[a_3 \times a_2] \cdot [a_3 \times a_4]}{|a_3 \times a_2||a_3 \times a_4|}\right),$$

where the sign coincides with the sign of $([a_2 \times a_4] \cdot a_3)$.

The hydrogen bonds between N_i and CO_j are modeled as a thin square well interaction of maximal distance $b_{max} = 4.2$ Å and minimal distance $b_{min} = 4.0$ Å with negative potential energy $-\epsilon_{HB}$. To maintain a correct orientation of the hydrogen bond, we introduce four auxiliary bonds which appear and disappear together with the hydrogen bond (Fig. 6a). These bonds are created between the reacting beads and their neighbors in the opposite backbone: $N_iC_{\alpha,j} = r_1$, $N_iN_{j+1} = r_2$, $CO_jC_{\alpha,i} = r_3$, and $CO_jCO_{i-1} = r_4$. The energy of these bonds is determined as a step function:

$$U(r_k) = \begin{cases} \infty & r_k < d_{min,k} \\ \epsilon_{HB} & d_{min,k} < r_k < d_{0,k}, \\ \epsilon_{HB}/2 & d_{0,k} < r_k < d_{1,k}, \\ 0 & d_{1,k} < r_k < d_{max,k}, \\ \infty & r_k > d_{max,k}. \end{cases} \tag{1}$$

The values of $d_{min,k}$, $d_{0,k}$, $d_{1,k}$, and $d_{max,k}$ are within the range of 4.4 Å–5.6 Å and their tables are presented in [87] The total potential energy change when N_i and CO_j come to a distance b_{max} is thus $\Delta U = -\epsilon_{HB} + \sum_{k=1}^{4} U(r_k)$. Obviously, $-\epsilon_{HB} < \Delta U < 3\epsilon_{HB}$. If one of the reacting beads already has a hydrogen bond, or if $|i - j| < 4$, or if the kinetic energy of N_i and CO_j is not sufficient to overcome the potential barrier ΔU, the hydrogen bond does not form and the beads collide as hard spheres. Otherwise, the hydrogen bond forms and the kinetic energy of the two beads is changed by $-\Delta U$. After the hydrogen bond has formed the molecular dynamics proceeds according to the general rules taking into account the discontinuities of the auxiliary bond potential until the beads N_i and CO_j again come at the distance b_{max}. At this point, the change in potential energy $\Delta U' = \epsilon_{HB} - \sum_{k=1}^{4} U(r_k)$ is computed and the hydrogen bond breaks if the kinetic energy of N_i and CO_j is sufficient to overcome the barrier $\Delta U'$. Thus during hydrogen-bond formation and breaking the total energy and momentum are strictly conserved. This algorithm results in a rather flexible hydrogen bond which can form if one of the reacting beads (e.g., CO) comes at any point on the surface of spherical segments of radius b_{max} surrounding another

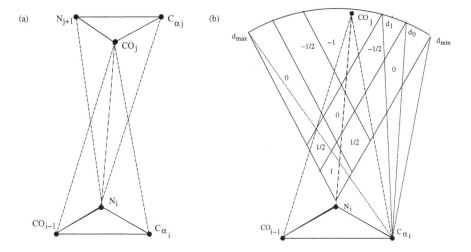

Fig. 6 (**a**) A hydrogen bond in our four-bead model. *Thick dashed lines* indicate hydrogen bond, *thin dashed lines* indicate auxiliary bonds helping to maintain hydrogen bond orientation, and *bold lines* show the covalent bonds of the backbone. See explanation in the text. (**b**) A projection of the potential energy landscape of the hydrogen bond onto the peptide plate. The *arc* indicates the sphere of radius b_{max}. Whenever a CO_j bead touches this sphere within spherical segments indicated by *thin bold lines* a hydrogen bond may form. The borders of these segments appear as *straight lines* on the projection because they are produced by the intersections of the hydrogen bond sphere and the auxiliary bond spheres of various radii ($d_{min} < d_0 < d_1 < d_{max}$, dotted lines) with the centers at the neighboring backbone beads CO_{i-1} and $C_{\alpha,i}$ which all lie in the projection plane. The numbers show the potential energy change according to (1) upon possible formation of this bond provided that the backbone of CO_j bead (not shown) has an optimal orientation. The drawing approximately reproduces the geometry for the values of d_{min}, d_0, d_1, and d_{max} given in [87]. The similar construction must be done for the contribution of the other two auxiliary bonds

bead (e.g., N) as indicated in Fig. 6b. These spherical segments are formed by the intersection of the sphere of radius b_{max} with the center at CO and the spheres of radii d_{min}, d_0, d_1, and d_{max} surrounding the neighboring beads (e.g., CO and C_α). Figure 6b shows the projection of these segments on the plane of beads $C_\alpha CON$, thus the boarders of these segments appear as straight lines. The numbers on the figure indicate the change in potential energy without taking into account the other two auxiliary bonds which are not shown $[\Delta U - U(r_1) - U(r_2)]$. This construction which belongs to Feng Ding [87] is the main difference between our four-bead model and the PRIME model of Carol Hall and co-workers [86].

We model a polyalanine of 16 amino-acids. The graphs of the Ramachandran angles for α-helix, β-hairpin, and random coil are in good agreement with the experimental ones (Fig. 7). The ground state of this model is the α-helix with potential energy $U_\alpha = -12\epsilon_{HB}$, which is significantly below the energy $U_\beta = -6\epsilon_{HB}$ of the β-hairpin; however the entropy S_α of the α-helix is very small comparatively to the entropy U_β of the β-hairpin which is visually apparent from the spread of the points on the Ramachandran plots. An accurate method of finding the entropy of the

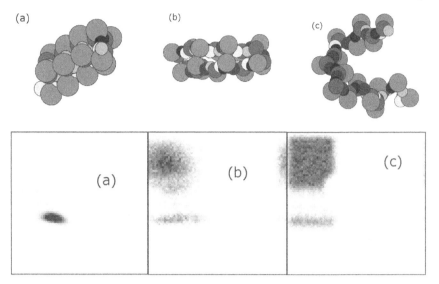

Fig. 7 *Upper panel*: typical conformations of the α-helix (**a**), β-hairpin (**b**), and random coil (**c**). The large *gray* beads represent C_β. The *gray* scale code indicates from black to white: N, CO, C_α, C_β, hydrogen bonded and terminal N, hydrogen bonded and terminal CO. *Lower panel*: Ramachandran plots for the α-helix (**a**), β-hairpin (**b**), and random coil (**c**) states of the 4-bead model. The scales range from 0 to 360° for both Φ (horizontal axis) and Ψ (vertical axis). Each cell corresponds to a bin of 5° in Φ and Ψ. Colors in rainbow order indicate the probabilities for Φ and Ψ to belong to a certain bin from black (*high*) to white (*low*). White cells indicate zero probability

4-bead model based on the root mean square deviation of the beads from a representative conformation is presented in our original publication [87]. The potential energy of the random coil is $U_c = 0$, but its entropy S_c is even higher. We found that there is a window of temperatures when the free energies $F_x = U_x - TS_x$ of all the three states are approximately equal. In fact, at $T = 0.12\epsilon_{HB}/k_B$, the free energies of β-hairpin and α-helix are populated with equal probability and the model can spontaneously undergo a reversible α-helix to β-hairpin transition (Fig. 8). We found that the only pathway from an α-helix to a β-hairpin leads through a completely unfolded conformation. The frequency of these transitions are proportional to $\exp((F_\alpha - F_c)/k_B T)$ and $\exp((F_\beta - F_c)/k_B T)$, which are highly dependent on temperature. At $T = 0.13\epsilon_{HB}/k_B$, the β-hairpin is at equilibrium with the random coil and the α-helix is almost never observed. In contrast, at $T = 0.11\epsilon_{HB}/k_B$, the spontaneous transition between an α-helix and a β-hairpin is never observed and the peptide can be trapped in a metastable β-hairpin conformation. If we estimate $\epsilon_{HB} = 21$ kJ/mol, [88] this temperature range corresponds to 276–328 K, i.e. to physiological conditions. Note that the transitions between all the three states resemble the first-order phase transitions. On the other hand, the β-hairpin can be regarded as a high temperature intermediate in the folding transition to the α-helix.

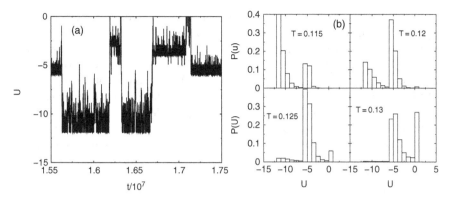

Fig. 8 (**a**) Potential energy versus time for the four-bead model of the 16-poly-alaninine peptide at $T = 0.125$ which is close to the equilibrium temperature of the β-hairpin ($U = 6\epsilon_{HB}$) to α-helix transition ($U = 12\epsilon_{HB}$). Several spontaneous transitions from a β-hairpin to an α-helix and back are observed. The transitions occur through complete unfolding to a random coil conformation ($U = 0$). The temperature window for which this behavior is possible is narrow. (**b**) Potential energy histograms for four different temperatures close to the glass transition. For $T = 0.115$, the random coil conformations are almost never observed, and the spontaneous transition becomes impossible in the simulation time scale window for $T < 0.11$. For $T = 0.13$ the peak of the histogram corresponding to the α-helix becomes almost invisible and for higher temperatures the transition to an α-helix never occurs

In order to study if the cooperative folding to the α-helix can be achieved by introducing side chain interactions, we include the hydrophobic interactions (H) for some of the C_β beads as square well potentials with depth ϵ_{HP} and range of attraction 6.5 Å. We model the polar side chains (P) by the hard spheres of the original diameter. We study the sequence PPHPPHHPPHPPHHPP, which was experimentally designed [86, 89, 90] to fold into the native α-helix. Indeed, when the relative strength of the hydrophobic interactions reaches $\epsilon_{HP} = 0.25\epsilon_{HB}$ we achieve the cooperative folding from a random coil to an α-helix, via a structureless molten globule state [66]. The folding transition is characterized by a very sharp maximum in the specific heat over the range of temperatures from 0.122 to 0.134. Thus, our model provides one of the first successful simulation of folding of a short peptide.

A missing link in understanding the amyloidogenesis of α-helix-rich proteins to β-sheet-rich fibrils is the possible presence of a metastable β-hairpin intermediate state, prone to aggregation [85]. Our results suggest a generic framework that explains why this β-hairpin intermediate is favorable in terms of free energy. Although the potential energy of a β-hairpin is higher than that of an α-helix, its entropy is also higher, thus the β-hairpin can appear as a high temperature intermediate. Also our simulation is consistent with the recent experimental results which show that the changes in the solvent can induce conformational changes in the protein. Indeed, the solvent can strongly affect the relative strength of hydrophobic interactions and hydrogen bonds. In recent years, a major progress in the DMD simulations of polyalanine and polyglutamine has been achieved by Hall and co-workers [84]. They also successfully simulated the effect of side chain interactions.

In [91] the four-bead model has been applied to the studies of the β-amyloid aggregation. The side chains of the amyloid peptide with 40(42) residues DAE-FRHDSGYEVHHQKLVFFAEDVGSNKGAIIGLMVGGVV(IA) have been represented by hard spheres located at C_β, except for six glycines (G), which have been presented by only three beads. At low $T = 0.1$ the conformations are mostly α-helical, while for larger T, mostly β-stranded conformations have been observed in agreement with the simulation of polyalanine [87]. When two identical peptides are placed in the simulation box they form various β-stranded dimers linked together by parallel or antiparallel β-strands. These conformations serve as initial conformations for the all-atom MD simulations which employ the Sigma program [92] for calculation of the free energy [93] difference to determine the stability of various dimer conformations and to compare the stability of $A\beta$-42 and $A\beta$-40 dimers. The majority of the peptide conformations produced by DMD passed the stability test by the all-atom MD. This fact suggests that the four-bead model generates realistic protein conformations. However, the observed conformations do not correspond to the structure of the $A\beta$ conformations in the $A\beta$ fibrils [94] and do not show significant differences between the stabilities of $A\beta$-42 and $A\beta$-40 dimers. This is not strange since the four-bead model used in these studies does not take into account the amino acid–specific interactions.

Later, the four-bead model has been used to investigate the aggregation of microglobulin [95], which is the molecular basis of the complications in patients undergoing long-term hemodialysis. These studies have addressed an important practical question of the role of the disulfide bond in the aggregation pathways. The side chain interactions have been modeled by the Go interaction matrix as in [79]. The resulting trimers produced by DMD have been further studied by all-atom AMBER-8 simulations [43].

9 Simulations of Amino Acid–Specific Interactions

The example provided by the four-bead model shows that we indeed have some hope to simulate ab initio folding of short proteins and make verifiable predictions of the protein aggregation and missfolding. Also, this example shows an extreme complexity of parameterization of the DMD model, with lots of parameters which do not have much physical meaning like auxiliary interactions used in the hydrogen-bond algorithm. Another problem is that very few research groups are working on the development of the DMD code and no well-documented source codes for the DMD simulations of biomolecules are publicly available. Obviously, development of such a code requires huge funding, while the granting agencies are focused on immediate biomedical applications and are reluctant to invest money in untraditional methods like DMD.

Nevertheless, some success has been achieved in this direction. In 2005, Ding et al. successfully simulate a folding transition of a trp-cage miniprotein NLY-IQWLKDGGPSSGRPPPS to a 1.5 Å resolution [61]. They have developed a more detailed "5 + +"-bead model of a protein with explicit carbonyl oxygens and heavy side chains with C_γ and C_δ beads, bifurcated side chains with two C_γ beads, and with an additional covalent bond between the side chain and the backbone for proline. The explicit carbonyl oxygens lead to a much more realistic and simple model of the backbone hydrogen bonds than in the four-bead model. Different type of interactions have been included for salt bridges, aromatic, and aromatic–proline interactions. They have found a folding transition with an intermediate of 3.5 Å from the native state and the folded state of 1.5 Å from the NMR-resolved native state. However, as they decrease the temperature, the intermediate does not completely disappear and the distance from the NMR native state increases. This discrepancy clearly indicates that the parameters of the model are still imperfect although all the interactions have been defined from the extensive statistical analysis of the protein structures in the PDB. Another concern is that trp-cage miniprotein is a very specific one with especially rigid backbone due to four prolines. Thus, the relative success in its folding may not be transferable to other proteins.

More recently, similar model [96] has been applied for the studies of polyglutamine aggregation, which is the cause of nine human diseases including Huntington's disease. Glutamine has been modeled by six beads: four beads of the backbone C_α, C', O, N and two beads of the side chain, one bead representing methylene groups ($-CH_2-CH_2-$) and another representing the carboxylamine ($CONH_2$) group. The authors show that the propensity to form β-sheets increases with the length of the polyglutamine repeat. This may explain why if the repeat length exceeds a critical value of 35–40 glutamines the disease starts to develop and becomes more severe as the repeat length increases in a lifetime of a patient. The same model have been also employed to study the α-helix-to-β-sheet transition with a subsequent aggregation of a 17-residue peptide named ccβ, SIRELEARIRELELRIG [97]. Conformational changes of this small protein-like peptide can serve as a model for prion diseases.

Recent DMD studies [98, 99, 100] have been done in close collaboration with the experimental group of David Teplow and are aimed to model specific pathways of amyloid aggregation in Alzheimer's disease [101, 102]. Urbanc et al. use DMD simulations of β-amyloid aggregation using the original four-bead model with various strength of hydrophobic interactions of the C_β beads, which are intended to model the amino acid–specific interactions. The hydrophobic strength has been taken from standard hydrophobicity tables [103]. These simulations have been done with 32 peptides and show that they aggregate into micelle-like disordered oligomers of various sizes with highly hydrophobic amino acids in the center and hydrophilic amino acids forming a shell around the hydrophobic core which prevents further aggregation. Interestingly, the Aβ-42 which is genetically linked with the Alzheimer disease phenotype forms larger oligomers than Aβ-40 which lacks a highly hydrophobic isoleucine at the C-terminus. This finding is constant with the experimental results of [81, 104]. These studies also reveal a statistically predominant turn near the N-terminus of the aggregated peptides. This turn is due to glycines Gly37 and Gly38 and is stabilized by the strong hydrophobic interaction between two valines Val36 and Val39. While in Aβ40, no other nearby amino acid is involved in this turn, in Aβ42, methionine Met35 strongly interacts with isoleucine Ile41 and valines Val39 and Val40 and thus stabilizes this turn even more. This is especially significant, since in vitro oxidation of methionine in Aβ42 by Bitan et al. [105] reduces the aggregation propensity of Aβ42 and makes it equal to that of Aβ-40.

An attempt to build a united atom DMD model has been recently made by Borreguero et al. [106]. This model is a further development of Feng et al. [61] 5 + +-bead model and takes into account all atoms except hydrogens. In collaboration with Teplow's group, they studied the conformational statistics of the central amyloid segment Aβ(21-31) which is believed to form a folded structure in the oligomers. The results of simulations predict a loop structure with a turn caused by the hydrophobic interactions between valine and a long hydrophobic part of lysine. The charged tail of lysine is competing to form a salt bridge with the aspartic acid and the glutamic acid. This salt bridge stabilizes the loop. By varying the strength of electrostatic interactions, it is possible to shift the most predominant electrostatic interaction from Glu22-Lys28 to Asp23-Lys28. This may indicate a special role which Glu22 has in familial mutations which increase the risk of Alzheimer disease. This example shows what level of molecular details can be achieved by DMD. Of course, all these predictions may not be correct because the behavior of the model highly depends on the hundreds of parameters describing the interactions. These parameters are obtained by Borreguero from the extensive studies of the PDB but still a huge effort on optimizing the interactions is needed before DMD will achieve predictive power. At present, the results of DMD can provide a useful food for thought for the experimentalists but one has to always take them with reservations. Nevertheless, I believe that with the increase of computation power and with enormous effort of devoted graduate students, DMD will soon become a routinely used predictive tool in molecular biology.

Acknowledgements I am very grateful to all the coauthors of all the joint works on DMD in which I have participated. I am obliged to A. Yu. Grosberg and A. R. Khokhlov who invited me in 1995 to make a CD for the popular book [27] *Giant Molecules*. This project led to the development of the fast and universal DMD code suitable for simulations of protein folding. I am grateful to E.I. Shakhnovich without whose expertise this work on protein folding and aggregation would be impossible. A crucial role in application of DMD particularly to protein folding belongs to the former BU graduate students N. V. Dokholyan, F. Ding, and J. Borreguero. They developed the two-bead, four-bead, and united atom models for DMD. In particular, F. Ding made significant additions to the DMD code for the hydrogen bonding. I thank Brigita Urbanc and Simcha Rimler for critical comments on the manuscript. Last but not least, I want to emphasize the role of H. E. Stanley whose interest in interdisciplinary studies made this collaboration possible. I acknowledge financial support of NSF, NIH, PRF, and anonymous private foundations. I also gratefully acknowledge the partial support of this research through the Dr. Bernard W. Gamson Computational Laboratory at Yeshiva College.

Appendix A Details of the DMD Algorithm

The structure of the DMD algorithm is the following:

(1) Find collision times of all pairs of neighboring particles and record them into collision tables
(2) Find the next collision time.

(3) Clean up the tables from the data involving colliding particles.

(4) Move the colliding pair to the time of collision and find the new velocities after collision.

(5) Find the new collision times of the collided particles with their neighbors.

(6) Compute time averages of the properties of interest and save the data if needed.

(7) Go to Step (2)

A.1 Find the Next Collision

Between collisions, particles move along straight lines with constant velocities. When the distance between the particles, r, becomes equal to r_k at which $U(r)$ has a discontinuity, the velocities of the interacting particles instantaneously change. The interaction time t_{ij} for two particles with coordinates r_i, r_j and velocities v_i, v_j satisfies the quadratic equation

$$(r_{ij} + t_{ij}v_{ij})^2 = R_{ij}^2,$$

where $R_{ij} = r_k$ and k depends on the initial distance between particles $r_{ij} = r_i - r_j$ and their relative velocity $v_{ij} = v_i - v_j$. This quadratic equation may have two positive roots, two negative roots, two roots of different signs, or no roots at all. The roots are determined by the formula

$$t_{ij} = \frac{-(v_{ij} \cdot r_{ij}) \pm \sqrt{(v_{ij} \cdot r_{ij})^2 + v_{ij}^2 \left(R_{ij}^2 - r_{ij}^2 \right)}}{v_{ij}^2},$$

where the actual collision time corresponds to sign "plus" if roots have different signs or "minus" otherwise. The value of k in $R_{ij} = r_k$ is selected to minimize $t_{ij} > 0$. If there are no positive roots, it means that the particles will not interact and $t_{ij} = \infty$.

A.2 Move the Colliding Particles Forward Until a Collision Occurs

We find the next collision time

$$\delta t = \min_{i<j} t_{ij}$$

for all possible pairs of particles and propagate the system to time

$$t' = t + \delta t$$

so that

$$r'_i = r_i + \delta t v_i.$$

At this moment, the distance between the centers of colliding particle-pairs becomes equal to r_k. The minimization of the t_{ij} is optimized by dividing the system into small cubic cells whose size is equal to the largest interaction distance. The collision times are computed only for particles in the nearest neighboring cells and are stored in the collision lists of the atoms belonging to this cell. After two particles collide, their future collision times with other atoms become invalid and must be removed from their collision lists and from the collision lists of the atoms in whose collision lists these particles occur. Only these affected atoms are moved to the next collision time, and the new collision times of these atoms with the collided particles are computed. The rest of the atoms in the system are not affected and stay at their positions.

In order to keep track of the atom position in space-time, each atom structure stores (besides the atom type, current coordinates, and velocity components) the update time, (i.e., the time at which the atom coordinates were last updated) and the time of leaving the cell. The collision times larger than this value are not kept. After an atom enters a new cell (which is treated as an event equivalent to the collision with other particles) its collision list is empty, and it is filled again with collisions with atoms in the new neighboring cells. After collision tables are updated, the new nearest collision time is found for each cell containing the affected atoms and then those cells participate in the binary tree sorting procedure similar to the World Cup schedule. This algorithm reduces the computational costs to $N \ln N$, where $\ln N$ comes from the tree sorting and for all practical purposes can be neglected. However, this algorithm becomes impractical when the largest interaction distance becomes greater than $1/4$ of the system box. Further improvements can be achieved computing a list of all atoms within certain distance to a given atom [33]. The program spends most of the CPU time in the calculation of the next collision times, many of which will never occur, because an atom will collide with somebody else sooner. The calculations of the future collision times during the update of the collision tables can be parallelized. But the problem of effective scalable parallelization for the DMD (to the best of my knowledge) has not yet been solved.

A.3 Implement Collision Dynamics of the Colliding Pair

Finally, we find the new velocities v'_i and v'_j after the collision. These velocities must satisfy the momentum conservation law

$$m_i v_i + m_j v_j = m_i v'_i + m_j v'_j,$$

the angular momentum conservation law

$$m_i [r'_i \times v_i] + m_j [r'_j \times v_j] = m_i [r'_i \times v'_i] + m_j [r'_j \times v'_j],$$

and the energy conservation law

$$\frac{m_i v_i^2}{2} + \frac{m_j v_j^2}{2} + U_{ij} = \frac{m_i v_i'^2}{2} + \frac{m_j v_j'^2}{2} + U_{ij}',$$

where U_{ij} and U_{ij}' are the values of the pair potential before and after the collision. These values are equal to $U(R_{ij} \pm \epsilon)$, depending on the direction of the initial relative velocity v_{ij}, initial distance r_{ij}, and the value of R_{ij}. These equations are equivalent to six scalar equations, which are sufficient to determine the six unknown components of the velocities v'_i and v'_j. By introducing a new coordinate system with the origin at the center of the particle j, and the x-axis collinear with the vector r'_{ij}, we construct the expressions for the velocities that satisfy the momentum and the angular momentum conservation laws:

$$\begin{aligned} v'_i &= v_i + A r'_{ij} m_j, \\ v'_j &= v_j - A r'_{ij} m_i, \end{aligned} \tag{2}$$

where constant A is determined from the energy conservation law:

$$A = a \frac{\pm \sqrt{1 + 2(U_{ij} - U_{ij}')(m_i + m_j)/(R_{ij}^2 a^2 m_i m_j)} - 1}{m_i + m_j} \tag{3}$$

and $a = (v_{ij} \cdot r_{ij})/R_{ij}^2$. The sign "plus" in the expression for A corresponds to the motion after the collision in the same direction as before the collision, i.e. the particles penetrate into the attractive well or the soft core if they move toward each other before the collision, or leave them if they move away from each other. Note that this may happen only if the expression under the square root is positive, i.e. if there is enough kinetic energy to overcome the potential barrier:

$$\frac{R_{ij}^2 a^2 m_i m_j}{2(m_i + m_j)} \geq U_{ij}' - U_{ij}.$$

Otherwise, the reflection happens, the particles do not change their state: $U_{ij}' = U_{ij}$, and the sign in the expression for A must be "minus".

Appendix B Calculation of Energy and Temperature

The total energy of our system is defined as:

$$E = K + U, \tag{4}$$

where K and U are the kinetic and potential energy, respectively. The kinetic energy is a sum of contributions from the individual particles

$$K = \sum_{i=1}^{N} \frac{m_i v_i^2}{2}, \tag{5}$$

while the evaluation of the potential energy contribution involves summation over all pairs of interacting particles

$$U = \sum_{i<j} U_{ij} . \tag{6}$$

The temperature of the system T is calculated according to the equipartition theorem. For a d-dimensional system the instantaneous temperature can be defined as

$$T = 2K/(Ndk_B) \tag{7}$$

B.1 Temperature Rescaling

In the DMD algorithm, the energy strictly conserves, so it corresponds to the microcanonical ensemble. To maintain the temperature constant or to slowly cool the system down, we use the Berendsen method [107] of velocity rescaling, multiplying all the velocities by a factor $\sqrt{T'/T}$, which is determined by

$$T' = T(1 - \kappa\Delta t) + \kappa\Delta t T_o , \tag{8}$$

where Δt is an approximately constant interval of time between two successive rescalings, T_o is the temperature of the heat bath, T is the instantaneous temperature before rescaling, T' is the instantaneous temperature after rescaling, and κ is the heat exchange coefficient. Usually, we select Δt as a time during which N collisions occur. In order to keep the old collision tables after rescaling, we actually rescale the energies of interactions and take this into account by keeping track of the ratio of the actual physical velocities and the unrescaled velocities in the computer. The inverse correction factor is applied to time. Interestingly, this correction factor exponentially inflates and may soon reach astronomical values. Once the correction factor becomes too large (e.g., 10) or too small (e.g. 0.1), we rescale velocities and time, return the interaction energies to the original value, and recalculate the collision tables from scratch.

Appendix C Calculation of Pressure

For ergodic systems, a thermodynamic average of a quantity f can be achieved in MD by averaging over a sufficiently large time Δt,

$$\langle f \rangle_{\Delta t} \equiv \frac{1}{\Delta t} \int_t^{t+\Delta t} f(t)\mathrm{d}t . \tag{9}$$

The calculation of pressure in MD has another difficulty because due to the periodic boundaries the system does not have walls which create external pressure.

Nevertheless, the average pressure P over a long enough period of time can be effectively computed using the virial theorem:

$$P = \frac{2}{Vd} \left\langle \sum_{i=1}^{N} \frac{m_i v_i^2}{2} \right\rangle_{\Delta t} + \frac{1}{Vd} \left\langle \sum_{i=1}^{N} f_i \cdot r_i \right\rangle_{\Delta t} \tag{10}$$

where f_i is the force acting on particle i from all other particles. Note that $\sum_{i=1}^{N} m_i v_i^2/2$ is by definition (7) equal to $dk_B NT/2$ and

$$P = \frac{Nk_B}{V} \langle T \rangle_{\Delta t} + \frac{1}{Vd} \left\langle \sum_{i=1}^{N} f_i \cdot r_i \right\rangle_{\Delta t} . \tag{11}$$

When the system has walls, this equation gives the value of the pressure acting from the walls to the system. In the absence of walls, it gives the value of the internal pressure in the system. Thus, this equation provides the basis for the computation of pressure in molecular dynamics simulations.

In discrete molecular dynamics, the force f_i is equal to zero except at the moments of collision with other particles, when it is equal to infinity. We count all the collisions of a given particle i with a given particle j that occur in the time interval from t to $t + \Delta t$, using index $K_{ij} = 1, 2, 3, \ldots$ We denote the times of these collisions $t_{K_{ij}}$ and the change in momentum of particle i at the moments $t_{K_{ij}}$ as

$$\Delta p_{K_{ij}} = m_i [v_i(t_{K_{ij}} + \epsilon) - v_i(t_{K_{ij}} - \epsilon)] , \tag{12}$$

where ϵ is an infinitesimally small value. Since the force acting on the particle i is the derivative of momentum with respect to time,

$$f_i = \sum_{j=1}^{N} \sum_{K_{ij}} \Delta p_{K_{ij}} \delta(t - t_{K_{ij}}) , \tag{13}$$

where $\delta(t - t_{K_{ij}})$ is a Dirac δ-function and the sum over K_{ij} is taken over all collisions between particle i and j during time interval $(t, t + \Delta t)$.

Integration involved in the averaging over time [see (9)] eliminates δ-functions and we obtain

$$P = \frac{Nk_B}{V} \langle T \rangle_{\Delta t} + \frac{1}{\Delta t V d} \sum_{i=1}^{N} \sum_{K_{ij}} \sum_{j=1}^{N} (\Delta p_{K_{ij}} \cdot r_i) . \tag{14}$$

Finally, we can count all the collisions that occur in interval $(t, t + \Delta t)$ by index ℓ. Each collision is specified by the particles $i(\ell)$ and $j(\ell)$ involved in the collision $(i < j)$ and is counted twice in the sum of (14)—the first time when i is from the first sum and the second when i is from the last sum. According to momentum conservation, $\Delta p_{i(\ell)} = -\Delta p_{j(\ell)}$. Thus we rewrite (14) as

$$P = \frac{Nk_B}{V} \langle T \rangle_{\Delta t} + \frac{1}{\Delta t V d} \sum_{\ell} \{\Delta p_{i(\ell)}(t_\ell) \cdot [r_i(t_\ell) - r_j(t_\ell)]\} , \tag{15}$$

where the sum is taken over all collisions ℓ that occur at moments t_ℓ during the time interval Δt. Finally, taking into account Eqs. (2) and (3),

$$P = \frac{Nk_B}{V} \langle T \rangle_{\Delta t} + \frac{1}{\Delta t V d} \sum_\ell m_{i(\ell)} m_{j(\ell)} R^2_{ij(\ell)} A_\ell, \tag{16}$$

where A_ℓ is given by expression (3) for $i = i(\ell)$, $j = j(\ell)$.

The DMD algorithm allows also the constant pressure simulations. The easiest way to do it is to apply Berendsen barostat, analogous to Berendsen thermostat, with the difference that now all the coordinates are rescaled periodically by a factor $(1 + \eta)$, where $\eta < \epsilon$ is a small quantity, proportional to the difference between the average pressure over this period of time and the desired pressure P_0 of the barostat. The problem is that after the rescaling some pairs of particles may occur in the zone of the infinite potential, since after rescaling they may become closer than their hardcore interaction distance or outside the range of the permanent bond. To solve this problem, we add an inner hard-core which constitutes $1 - \epsilon$ of the true hard core, and the outer bond distance which is larger than actual bond distance by factor of $1 + \epsilon$. The particles that appear to be within the gap between the actual and the inner hardcore cannot go inside the inner hardcore but can freely move through the outer hardcore. The analogous algorithm works for the bonds. No new rescaling takes place before all the particles become outside the actual hardcore. Of course, after rescaling, all the collision tables must be reconstructed from scratch. So rescaling should not be done too often, otherwise the simulation will significantly slow down.

References

1. A. Smith: Nature **426**, 883 (2003)
2. C. M. Dobson: Nature **426**, 884 (2003)
3. D. J. Selkoe: Nature **426**, 900 (2003)
4. B. J. Alder, T. E. Wainwright: J. Chem. Phys. **31**, 459 (1959)
5. M. R. Sadr-Lahijany, A. Scala, S. V. Buldyrev, H. E. Stanley: Phys. Rev. Lett. **81**, 4895 (1998)
6. G. Franzese, G. Malescio, A. Skibinsky, S. V. Buldyrev, H. E. Stanley: Nature **409**, 692 (2001)
7. S. V. Buldyrev, H. E. Stanley: Physica A **330**, 124 (2003)
8. A. Skibinsky, S. V. Buldyrev, G. Franzese, G. Malescio, H. E. Stanley: Phys. Rev. E **69**, 61206 (2004)
9. P. Kumar, S. V. Buldyrev, F. Sciortino, E. Zaccarelli, H. E. Stanley: Phys. Rev. E **72**, 021501 (2005)
10. Z. Yan, S. V. Buldyrev, N. Giovambattista, H. E. Stanley: Phys. Rev. Lett. **95**, 130604 (2005)
11. P. A. Netz, S. V. Buldyrev, M. C. Barbosa, H. E. Stanley: Phys. Rev. E **73**, 061504 (2006)
12. L. Xu, S. V. Buldyrev, C. A. Angell, H. E. Stanley: Phys. Rev. E **74**, 031108 (2006)
13. D. C. Rapaport: J. Phys. A **11**, L213 (1978)
14. D. C. Rapaport: J. Chem. Phys. **71**, 3299 (1979)
15. Y. Zhou, C. K. Hall, M. Karplus: Phys. Rev. Lett. **77**, 2822 (1996)
16. Y. Zhou, M. Karplus, J. M. Wichert, C. K. Hall: J. Chem. Phys. **107**, 10691 (1997).
17. S. W. Smith, C. K. Hall, B. D. Freeman: J. Comp. Phys. **134**, 16 (1997)
18. N. V. Dokholyan, E. Pitard. S. V. Buldyrev, H. E. Stanley: Phys. Rev. E **65**, R030801 (2002)

19. S. V. Buldyrev, P. Kumar, P. G. Debenedetti, P. J. Rossky, and H. E. Stanley, Proc. Natl. Acad. Sci. USA **101**, 21077 (2007).
20. G. Foffi, K. A. Dawson, S. V. Buldyrev, F. Sciortino, E. Zaccarelli, P. Tartaglia: Phys. Rev. E **65**, 050802 (2002)
21. E. Zaccarelli, S. V. Buldyrev, E. La Nave, A. J. Moreno, I. Saika-Voivod, F. Sciortino, P. Tartaglia: Phys. Rev. Lett. **94**, 218301 (2005)
22. A. J. Moreno, S. V. Buldyrev, E. La Nave, I. Saika-Voivod, F. Sciortino, P. Tartaglia, E. Zaccarelli: Phys. Rev. Lett. **95**, 157802 (2005)
23. C. Davis, H. Nie, N. V. Dokholyan: Phys. Rev. E, **75**, 051921 (2007)
24. S. Sharma, F. Ding, N. V. Dokholyan: Biophysical Journal, **92**, 1457 (2007)
25. F. Ding, N. V. Dokholyan: Trends Biotech. **23**, 450 (2005).
26. B. Urbanc, J. M. Borreguero, L. Cruz, H. E. Stanley: Methods in Enzymology **412**, 314 (2006).
27. A. Yu. Grosberg, A. R. Khokhlov: *Giant Molecules* (Academic Press, 1997)
28. Center for Polymer Studies, Boston University: Virtual Molecular Dynamics Laboratory, http://cps.bu.edu/education/vmdl/ (2007)
29. D. C. Rapaport: J. Comput. Phys. **34**, 184 (1980)
30. M. P. Allen, D. J. Tildesley, *Computer Simulation of Liquids* (Oxford University Press, New York, 1989)
31. D. C. Rapaport: *The Art of Molecular Dynamics Simulation* (Cambridge University Press: Cambridge, 1997)
32. D. Frenkel, B. Smit: *Understanding Molecular Simulation: From Algorithms to Applications* (Academic, San Diego, 1996)
33. A. Donev, S. Torquato, F. H. Stillinger: J. Comp. Phys. **202**, 737 (2005)
34. A. Donev, S. Torquato, F. H. Stillinger: J. Comp. Phys. **202**, 765 (2005)
35. C. De Michele, A. Scala, R. Schilling, F. Sciortino: J. Chem. Phys. **124**, 104509, (2006)
36. C. De Michele, S. Gabrielli, P. Tartaglia, F. Sciortino: J. Phys. Chem. B **110**, 8064 (2006)
37. Y. Zhou, M. Karplus: Proc. Natl. Acad. Sci. USA **94**, 14429 (1997)
38. E. A. Jagla: Phys. Rev. E **58**, 1478 (1998)
39. B. R. Brooks, R. E. Bruccoleri, B. D. Olafson, D. J. States, S. Swaminathan, M. Karplus: J. Comp. Chem. **4**, 187 (1983)
40. A. D. MacKerell, Jr., B. Brooks, C. L. Brooks, III, L. Nilsson, B. Roux, Y. Won, M. Karplus: CHARMM: The Energy Function and Its Parameterization with an Overview of the Program. In: *The Encyclopedia of Computational Chemistry*, vol. 1, ed. P. v. R. Schleyer et al. (John Wiley & Sons: Chichester, 1998) pp. 271–277
41. D. van der Spoel, E. Lindahl, B. Hess, G. Groenhof, A. E. Mark, H. J. C. Berendsen: J. Comp. Chem. **26**, 1701 (2005)
42. L. Kale, R. Skeel, M. Bhandarkar, R. Brunner, A. Gursoy, N. Krawetz, J. Phillips, A. Shinozaki, K. Varadarajan, K. Schulten: J. Comp. Phys. **151**, 283312 (1999)
43. D. A. Case, T. E. Cheatham, III, T. Darden, H. Gohlke, R. Luo, K. M. Merz Jr., A. Onufriev, C. Simmerling, B. Wang, R. Woods: J. Computat. Chem. **26**, 1668 (2005).
44. S. J. Plimpton: J. Comp. Phys. **117**, 1 (1995)
45. H. D. Nguyen, C. K. Hall: Proc. Natl. Acad. Sci. USA **101**, 16180 (2004)
46. M. Karplus: Fold. Des. **2**, 569 (1997)
47. E. I. Shakhnovich: Chem. Rev. **106**, 1559 (2006)
48. R. R. Shearer: Art Journal, **55**, 64 (1996).
49. K. B. Zeldovich, P. Chen, B. E. Shakhnovich, E. I. Shakhnovich: PLoS Comp. Bio. **7**, 1224 (2007)
50. S. Miyazawa, R. L. Jernigan: J. Mol. Biol. **256**, 623 (1996).
51. N. V. Dokholyan, S. V. Buldyrev, H. E. Stanley, E. I. Shakhnovich: Folding & Design **3**, 577 (1998).
52. H. Taketomi, Y. Ueda, N. Go: Int. J. Peptide Protein Res. **7**, 445 (1975)
53. N. Go, H. Abe: Biopolymers **20**, 991 (1981)
54. H. Abe, N. Go: Biopolymers **20**, 1013 (1981)

55. N. V. Dokholyan, S. V. Buldyrev, H. E. Stanley, E. I. Shakhnovich: J. Mol. Biol. **296**, 1183 (2000)
56. A. Scala, N. V. Dokholyan, S. V. Buldyrev, H. E. Stanley, E. I. Shakhnovich: Phys. Rev. E **63** 032901 (2001)
57. Y. Zhou, M. Karplus: Nature **401**, 400 (1999)
58. Y, Zhou, M. Karplus: J. Mol. Biol. **293**, 917 (1999)
59. V. S. Pande, I. Baker, J. Chapman, S. P. Elmer, S. Khaliq, S. M. Larson, Y. M. Rhee, M. R. Shirts, C. D. Snow, E.J. Sorin, B. Zagrovic: Biopolymers **68**, 91 (2003)
60. A. R. Fersht: Curr. Opin. Struc. Biol. **7**, 3 (1997)
61. F. Ding, S. V. Buldyrev, N. V. Dokholyan: Biophys. J. **88**, 147 (2005)
62. J. M. Borreguero, N. V. Dokholyan, S. V. Buldyrev, H. E. Stanley, E. I. Shakhnovich: J. Mol. Biol. **318**, 863 (2002)
63. F. Ding, W. Guo, N. V. Dokholyan, E. I. Shakhnovich, J.-E. Shea: J. Mol. Biol. **350**, 1035 (2005)
64. J. M. Borreguero, F. Ding, S. V. Buldyrev, H. E. Stanley, N. V. Dokholyan: Biophys. J. **87**, 521 (2004).
65. H. Jang, C. K. Hall, Y. Zhou: Biophys. J. **83**, 819 (2002)
66. O. B. Ptitsyn: Adv. Protein Chem. **47**, 83 (1995)
67. F. Ding, R. K. Jha, N. V. Dokholyan, Structure **13**, 1047 (2005)
68. K. Lum, D. Chandler, J. Weeks: J. Phys. Chem. B **103**, 4570 (1999).
69. F. Ding, N. V. Dokholyan, S. V. Buldyrev, H. E. Stanley, E. I. Shakhnovich: Biophys. J. **83**, 3525 (2002)
70. S. Sharma, F. Ding, H. Nie, D. Watson, A. Unnithan, J. Lopp, D. Pozefsky, N. V. Dokholyan: Bioinformatics **22**, 2693 (2006)
71. C. Clementi, H. Nymeyer, J. N. Onuchic, J. Mol. Biol. **298**, 937 (2000)
72. A. R. Lam, J. M. Borreguero, F. Ding, N. V. Dokholyan, S. V. Buldyrev, H. E. Stanley, E. Shakhnovich: J. Mol. Biol. **373**, 1348 (2007).
73. A. R. Fersht: Proc. Natl. Acad. Sci. USA. **97**, 14121 (2000)
74. Y. Chen, N. V. Dokholyan, J. Biol. Chem. **281**, 29148 (2006)
75. J. W. Kelly: Curr. Opin. Struct. Biol. **8**, 101 (1998)
76. S. Y. Tan, M. B. Pepys: Histopathology **25**, 403 (1994)
77. F. Ding, N. V. Dokholyan, S.V. Buldyrev, H. E. Stanley, E. I. Shakhnovich: J. Mol. Biol. **324**, 851 (2002)
78. F. Ding, K. C. Prutzman, S. L. Campbell, N. V. Dokholyan: Structure, **14**, 5 (2006).
79. S. Peng, F. Ding, B. Urbanc, S. V. Buldyrev, L. Cruz, H. E. Stanley, N. V. Dokholyan: Phys. Rev. E **69**, 041908 (2004)
80. D. M. Walsh, I. Klyubin, J. V. Fadeeva, W. K. Cullen, R. Anwyl, M. S. Wolfe, M. J. Rowan, D. J. Selkoe: Nature **416**, 535 (2002)
81. G. Bitan, M. D. Kirkitadze, A. Lomakin, S. S. Vollers, G. B. Benedek, D. B. Teplow: Proc. Natl. Acad. Sci. USA **100**, 330 (2003)
82. H. D. Nguyen, C. K. Hall: Biophys. J. **87**, 4122 (2004)
83. H. D. Nguyen, C. K. Hall: J. Biol. Chem. **280**, 9074 (2005)
84. A. J. Marchut, C. K. Hall: Biophys. J. **90**, 4574 (2006)
85. J. S. Richardson, D. C. Richardson: Proc. Natl. Acad. Sci. USA **99**, 2754 (2002)
86. A. V. Smith, C. K. Hall: Proteins Struct. Funct. Genet. **44**, 344 (2001)
87. F. Ding, J. M. Borreguero, S. V. Buldyrev, H. E. Stanley, N. V. Dokholyan: Proteins Struct. Funct. Genet. **53** 220 (2003)
88. B. Honig, A. S. Yang: Adv. Protein Chem. **46**, 2758 (1995).
89. S. P. Ho, W. F. DeGrado: J. Am. Chem. Soc. **109**, 6751 (1987)
90. Z. Guo, D. Thirumalai: J. Mol. Biol. **263**, 323 (1996)
91. B. Urbanc, L. Cruz, F. Ding, D. Sammond, S. Khare, S. V. Buldyrev, H. E. Stanley, N. V. Dokholyan: Biophys. J. **87** 2310 (2004)
92. J. Hermans, R. H. Yun, J. Leech, D. Cavanaugh: Sigma documentation, University of North Carolina (1994). http://hekto.med.unc.edu:8080/HERMANS/software/SIGMA/index.html

93. Y. N. Vorobjev, J. Hermans: Biophys. Chem. **78**, 195 (1999).
94. A. T. Petkova, Y. Ishii, J. J. Balbach, O. N. Antzutkin, R. D. Leapman, F. Delaglio, R. Tycko: Proc. Natl. Acad. Sci. USA **99**, 16742 (2002).
95. Y. Chen, N. V. Dokholyan: J. Mol. Biol. **354**, 473 (2005).
96. S. D. Khare, F. Ding, K. N. Gwanmesia, N. V. Dokholyan, PLoS Comp. Biol. **1**, e30 (2005).
97. F. Ding, J. J. LaRocque , N. V. Dokholyan, J. Biol. Chem. **280**, 40235 (2005).
98. B. Urbanc, L. Cruz, S. Yun, S. V. Buldyrev G. Bitan, D. B. Teplow, H. E. Stanley, "In Silico Study of Amyloid Beta Protein Folding and Oligomerization," Proc. Natl. Acad. Sci. **101**, 17345–17350 (2004).
99. S. Yun, B. Urbanc, L. Cruz, G. Bitan, D. B. Teplow, H. E. Stanley: Biophys. J. **92**, 4064 (2007).
100. A. Lam, B. Urbanc, J. M. Borreguero, N. D. Lazo, D. B. Teplow, H. E. Stanley: Discrete Molecular Dynamics Study of Alzheimer Amyloid β-protein, *Proceedings of The 2006 International Conference on Bioinformatics & Computational Biology*, CSREA Press, Las Vegas, Nevada, 322–328 (2006).
101. B. Urbanc, L. Cruz, D. B. Teplow, H. E. Stanley, Current Alzheimer Research **3**, 493 (2006).
102. D. B. Teplow, N. D. Lazo, G. Bitan, S. Bernstein, T. Wyttenbach, M. T. Bowers, A. Baumketner, J.-E. Shea, B. Urbanc, L. Cruz, J. Borreguero, H. E. Stanley: Account of Chemical Research **39**, 635 (2006).
103. J. Kyte, R. F. Doolittle: J. Mol. Biol. **157**, 105 (1982)
104. G. Bitan, A. Lomakin, D. B. Teplow: J. Biol. Chem. **276**, 35176 (2001)
105. G. Bitan, B. Tarus, S. S. Vollers, H. A. Lashuel, M. M. Condron, J. E. Straub, D. B. Teplow: J. Am. Chem. Soc. **125**, 15359 (2003)
106. J. M. Borreguero, B. Urbanc, N. D. Lazo, S. V. Buldyrev, D. B. Teplow, H. E. Stanley, Folding events in the 21-30 region of amyloid-beta-protein (A beta) studied in silico,6020 Proc. Natl. Acad. Sci. **102**, 6015 (2005)
107. H. J. C. Berendsen, J. P. M. Postma, W. F. van Gunsteren, A. Di Nola, J. R. Haak: J. Chem. Phys. **81**, 3684 (1984)

Cooperative Effects in Biological Suspensions: From Filaments to Propellers

I. Pagonabarraga and I. Llopis

Abstract We analyze the role of hydrodynamic cooperativity in different systems of biological interest. We describe alternative approaches to model the dynamics of suspensions which account for realistic dynamic coupling on the length and time sales in which mesoscopic suspended particles (such as semiflexible filaments and swimmers) evolve. The time evolution and transport of such systems, which are relevant in physical biology, are studied in detail.

We consider the interactions in the dynamic response of inextensible semiflexible filaments subject to uniform external drivings and describe the interplay between elastic and hydrodynamic stresses. We analyze how such couplings give rise to a rich phenomenology in their dynamics. Using a complementary mesoscopic approach, we also discuss the dynamic regimes of suspensions of self-propelling particles which interact only through the embedding solvent.

1 Introduction

Biological processes at the cellular level include in many instances the interplay of an embedding solvent. Inside the cell, citoplasmatic streaming [1] encompasses the convective motion of fluid which favors a more efficient transport of material inside the cell [2]. Also cargo transport through molecular motors or the networks of actin and microtubules in eukaryotic cells have a dynamic behavior analogous to a viscoelastic fluid [3].

At the cellular level, the motility of cells and other microorganisms takes place in a solvent whenever the displacement does not take place on contact to a solid substrate [4]. It has long been recognized that the sizes and typical fluid viscosities

I. Pagonabarraga
Departament de Física Fonamental, Universitat de Barcelona, Carrer Martí i Franqués 1, 08028-Barcelona, Spain, ipagonabarraga@ub.edu

I. Llopis
Departament de Física Fonamental, Universitat de Barcelona, Carrer Martí i Franqués 1, 08028-Barcelona, Spain, isaacll@ub.edu

Pagonabarraga, I., Llopis, I.: *Cooperative Effects in Biological Suspensions: From Filaments to Propellers*. Lect. Notes Phys. **752**, 133–152 (2008)
DOI 10.1007/978-3-540-78765-5_6   © Springer-Verlag Berlin Heidelberg 2008

imply that the characteristic Reynolds numbers associated to the flows generated by microorganism motion are extremely small, e.g. for a microorganism of 10μm size in a water solution (with kinematic viscosity $v \sim 10^{-2} cm^2/s$) moving at a characteristic speed $u \sim 10 \mu m/s$ has $Re = uR/v \sim 10^{-4}$. In this regime, inertia is irrelevant and propulsion is only feasible if the cyclic motion generated by the organisms is non-reciprocal [5]. This fact has been clearly recognized, and recently minimal models for propulsion have been proposed [6], although the interplay between simplicity and efficiency is still open to scrutiny.

The understanding of the molecular structure underlying many of the basic polymers and membranes which compose cells has also allowed to understand how rotating motors on the base of microtubules determine the origin of flagellar motion [7], or how cilia beat [8, 9]. The collective motion of such filaments has also started to be analyzed and it has been shown that the dynamic coupling through the underlying fluid favors the synchronization of cilia and the development of metachronal waves [10, 11].

In many of the described tasks, the organisms or involved elements consume energy to perform the appropriate activity. Such active processes can take place as a result of chemo-mechanical internal means (in many cases mediated by molecular motors) and therefore do not require the action of external forces. Such type of motion is peculiar and differs from the situation in which passive suspended polymers or colloids respond to the action of external forces. In the latter situation the motion of such objects decay algebraically at long distances as algebraically as the inverse of the distance to the moving particle, while for active motion the induced fluid flow decays algebraically but faster.

There is a need to develop and use mesoscopic methods which include the relevant dynamic couplings of biological active materials and the solvent they are in many cases in contact with. Since the pertinent objects are micrometric, and therefore orders of magnitude larger than the liquid molecules, the time and lengths scales which characterize their evolution are orders of magnitude larger than the corresponding ones for the solvent. Hence, it is not possible to resolve the dynamics of the degrees of freedom of the fluid using state-of-the-art numerical platforms. The knowledge developed in colloid and polymer physics has made it clear that suprananometric objects interact with a large number of molecules almost instantaneously and consequently the direct individual interactions do not play a relevant role in colloid motion. Rather, the suspended particles are sensitive to the collective modes of the underlying fluid. Hence, it is possible to devise mesoscopic theories which combine a detailed resolution of the structure and dynamics of the mesoscopic particles coupled to the fluid conserved variables (i.e., propose a dynamic theory for the solvent's density and momentum). Such approaches naturally adapt themselves to the scales in which the particles evolve and capture the relevant physical couplings. In biological fluids the embedding medium is in turn composed by many other particles and exhibits in general viscoelastic behavior. That information should be included in these models; hence, the properties of the liquid should be determined by other means.

We will describe two complementary approaches to the study of the dynamics and cooperative behavior of biologically relevant objects which are moving in a Newtonian solvent. We will describe the peculiarities that such couplings give rise to simplified geometries. Specifically, it is necessary the development of methods and numerical tools that capture the characteristics of biofilaments and which can also describe accurately the relevant dynamic couplings between the suspended objects and the fluid they move in. We will address these questions in the subsequent sections. Specifically, we will introduce a method that describes inextensible semiflexible polymers and which can serve as a basis to the study of dynamical processes of biofilaments and constitutes a simple model for flagella. Subsequently, we will introduce a simple model that captures self-propulsion for simple models of swimmers with complete hydrodynamic coupling.

2 Semiflexible Filaments

Semiflexible polymers, such as DNA, actin or microtubules, are abundant in biology, where they play multiple roles related to cellular mechanics and self-organization [4]. They are distinct from rigid rods or flexible polymers in that even if the conformational changes need to be accounted for, there is a memory of the filament's shape, which is usually described in terms of the filament persistence length, L_p [12]. Therefore, their physical properties differ from those of other types of polymers.

The internal, molecular, structure of biofilaments makes them essentially inextensible [4], as opposed to other semiflexible polymers, such as polyelectrolytes. The inextensible and semiflexible properties of these polymers confer them specific properties and qualitatively different responses to externally or internally applied forces and stresses. Understanding such specific behavior is relevant to assess, e.g., the mechanisms of flagellar locomotion or cilia beating. We are interested in analyzing the relevance of biopolymer peculiar elasticity on their motion, reaction to imposed stresses. We will concentrate on simplified geometries to highlight the different response of these polymers when compared with flexible and rigid polymers and analyze their peculiar hydrodynamic cooperativity.

3 Modeling Inextensible Semiflexible Filaments

A semiflexible filament suspended in a low Reynolds number fluid can be modeled as a discrete chain of N beads which evolve according to Newton's equations. The interaction with the fluid enters through dissipative, friction forces and accounts for the interactions between beads mediated by the fluid. Due to this friction, inertia is nor relevant in the motion of suspended chains. We insure that the inertial transients are negligible by choosing an appropriately small value of the beads' mass [13].

We describe now in turn the different forces that must be included in such a dynamical model for an inextensible semiflexible chain.

3.1 Bending Rigidity

A semiflexible filament has an elastic resistance to bending, described by its bending stiffness κ. In a continuum description, the bending elasticity is given by the Hamiltonian

$$\mathscr{H} = \frac{1}{2}\kappa \int_0^L C^2(s)\mathrm{d}s \ , \tag{1}$$

where s describes the position along the filament's arclength and $C(s)$ corresponds to the local curvature at s. In our model, the bending energy is expressed as the discretization of (1)

$$\mathscr{H}_b = \frac{1}{2}\kappa b \sum_{i=2}^{N-1} C_i^2 \ . \tag{2}$$

If θ_i is the angle formed by the bond that connects bead $i-1$ to bead i and the one that connects bead i with bead $i+1$, the local curvature can be written down as

$$C_i^2 = \frac{2}{b^2}(1 - \cos\theta_i), \tag{3}$$

and the bending force on bead i, $\mathbf{F}_{iB}$, is obtained as the variation of $\mathscr{H}$ when bead i changes its position.

3.2 Inextensibility

We impose that consecutive beads are connected by bonds of fixed length b such that the filament has a length $L = (N-1)b$, which does not vary with time. This is done through the efficient implementation of conveniently chosen constraint forces ($\mathbf{F}_{iC}$), as proposed earlier [13, 14]. Such a feature appears as a result of the strength of covalent bonds at microscopic scales [15].

3.3 Friction Forces and Hydrodynamic Interactions

We want to account for the presence of a suspending fluid and the force it exerts on the filament. To this end, we need to consider that the total force $\mathbf{F}_j$ acting on bead j generates a flow that will tend to drag the rest of the beads. The flow bead j generates can be described at the level of the Oseen approximation; the hydrodynamic velocity, $\mathbf{v}_i^H$, at a bead located at $\mathbf{r}_i$ induced by the forces acting on the rest of the beads can be written down as

$$\mathbf{v}_i^H(t) = \frac{3}{4\gamma_0}\frac{a}{b}\sum_{j\neq i}\frac{1+\mathbf{e}_{ij}(t)\mathbf{e}_{ij}(t)}{r_{ij}(t)/b}\cdot\mathbf{F}_j(t), \tag{4}$$

where a stands for the bead size and $\mathbf{e}_{ij} = (\mathbf{r}_i - \mathbf{r}_j)/r_{ij}$ is the unit vector joining beads i and j, $r_{ij} = |\mathbf{r}_i - \mathbf{r}_j|$ corresponding to their distance. If bead i does not follow the local fluid velocity, it will suffer a hydrodynamic friction force

$$\mathbf{F}_{iF} = -\gamma_0(\mathbf{v}_i - \mathbf{v}_i^H). \tag{5}$$

where γ_0 is a characteristic parameter related to the fluid viscosity. The inertial relaxation time scale depends on both m and γ_0, and hence m should be tuned appropriately once γ_0 is fixed.

This description allows to include the hydrodynamic interactions (HI) between the filament beads in terms of a friction force. In this way we are going beyond resistive force theory, which takes $\mathbf{v}_i^H = 0$ and regards the solvent as a passive medium [16]. Such a standard approach does not account for the fact that the effective friction suffered by a filament depends on its configuration.

3.4 Time Evolution

The description in terms of a local friction force allows to describe the filament dynamics using a molecular dynamics like approach based on the total force acting on each bead. At each time step, the total force is computed as $\mathbf{F}_i = \mathbf{F}_{iB} + \mathbf{F}_{iC} + \mathbf{F}_{iF} + \mathbf{F}_{ie} + \mathbf{F}_{ith}$, where $\mathbf{F}_{ith}$ is the random force, which accounts for thermal fluctuations, and $\mathbf{F}_{ie}$ refers to the applied external force. In the following we will concentrate on filament's sedimentation, where $\mathbf{F}_{ie} \equiv \mathbf{F}_e$ is a constant external force, in situations where the energetic contributions due to the elastic energy dominate and will therefore neglect thermal forces ($\mathbf{F}_{ith} = \mathbf{0}$).

We will implement the velocity Verlet algorithm [17], which can be expressed explicitly in terms of update rules for the velocities and positions of every bead at time $t + \Delta t$ in terms of their counterparts at the previous time step, t,

$$\mathbf{r}_i(t+\Delta t) = \mathbf{r}_i(t) + \mathbf{v}_i(t)\Delta t + \frac{\Delta t^2}{2m_i}\mathbf{F}_i(t) \tag{6}$$

$$\mathbf{v}_i(t+\Delta t) = \mathbf{v}_i(t) + \frac{\Delta t}{2m_i}\left[\mathbf{F}_i(t) + \mathbf{F}_i(t+\Delta t)\right] \tag{7}$$

In order to analyze in detail the interplay between elasticity and HI, we will focus in simple geometries that involve a few filaments under the action of different types of external forces.

4 Semiflexible Filaments Under External Forcing

A filament under the action of a uniform external field of magnitude F_e will reach a steady state characterized by a constant sedimentation velocity. Such a situation can be found if an electric field is applied, or if the solution is placed in an ultracentrifuge. The speed will depend both on the field strength and on the chain mobility; the latter is sensitive to the filament orientation.

An elastic filament will evolve into a random coil configuration; in this case HI imply that the mobility grows as $D \sim N^{-\nu}$, which indicates that the coil behaves dynamically as a soft sphere of a size given by the equilibrium radius of gyration of the polymer, which grows with polymerization as $R_g \sim bN^\nu$. The coil will keep its spherical shape until a large enough field is applied, when secondary blobs are observed and the chain undergoes a series of shape transitions [18].

Some biofilaments, such as the tobacco mosaic virus, behave as rigid rods. In this case particles will move at constant velocity when subjected to external forces. The rods move without changing their orientation. For a rod the velocity depends on the relative orientation between the rod and the applied force; the velocity is maximal for a rod aligned with the force, characterized by a parallel mobility $\mu_\parallel$ and has a minimum if the rod is transversed to the applied field, where its velocity is proportional to the applied force by a factor equal to its transverse mobility, $\mu_\perp$ (with $\mu_\parallel = 2\mu_\perp$ [12]). The random motion of the rods will make them explore the different possible orientations and in this case the steady state velocity will be characterized by a mean mobility which will average the perpendicular and parallel limiting values.

The situation of a semiflexible filament is different. In this case, numerical simulations using the previously introduced model show that for any initial condition, the filament rotates until it gets an orientation perpendicular to the applied field [20]. The conformation the filament develops depends on its bending rigidity. We display in Fig. 1 the deformation of the filament as a function of the dimensionless applied field, $B = L^3 F_e / \kappa$. This dimensionless parameter is a measure of the energy introduced by the external field on the filament's distance against the energy cost of deforming the filament. When this parameter exceeds unity, the filament deformation becomes of the order of the filament's size. Accordingly, one can identify a regime of weak deformation, $B < 200$, where the filament is slightly bent, and a high forcing regime, $B > 200$, where the filament has a saturated curved shape.

For real biopolymers, such as 10μm microtubules, $B < 10^{-3}$ in a gravitational field, and can go up to $B \sim 1$ in an ultracentrifuge. although larger values can be achieved under the action of electric fields, a situation of relevance in electrophoresis. A different kind of semiflexible polymers can be achieved by joining magnetic colloidal particles with DNA strands. The corresponding chains respond to applied magnetic fields. In these systems, bending and alignment of chains analogous to the ones discussed here are observed. In this case, the magnetic field exerts a torque on neighboring colloids which favors the deformation and subsequent alignment [19].

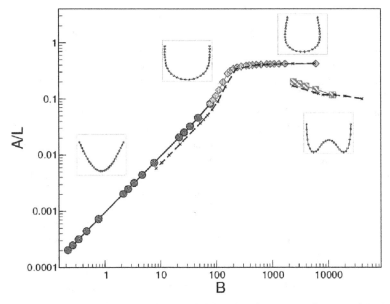

Fig. 1 Amplitude deformation of a single sedimenting filament as a function of the dimensionless parameter B. Insets: different filament configurations in the different regimes

The HI among filament beads lead to an unbalance in the hydrodynamic stress the filament is subject to. This is due to the fact that beads near the extremities of the filament feel a weaker friction than central ones due to the lower number of neighboring beads. As a result, the filament has to bend to balance the spatially non-uniform friction force that the fluid exerts on it. The sedimentation velocity in this case will depend on the degree of filament flexibility. For small B, it will correspond to $\mu_\perp$, and will increase as a function of B until the plateau region is reached. We have computed these frictions and have seen a decrease by a factor 2 [20].

If B is increased further, we have observed the development of metastable shapes (as displayed in Fig. 1) which relax eventually to a highly curved shape. At even larger values, the filament does not reach a steady state and becomes progressively spherical. One can hypothesize that the eventual shape becomes equivalent to that of a random coil corresponding to an elastic filament (one expects that the filament is equivalent to a flexible chain when the bending energy becomes irrelevant, compared to the external one, in the limit $1/B \rightarrow 0$). It is also then possible that the analogous shape transitions described in [18] are also observed here, outside the range of parameters sampled in our simulations.

Away from the infinite dilute limit, when several filaments sediment in the presence of each other, new effects arising from interchain HI are present. In this case elastic chains will behave as porous spheres and their relative distance during sedimentation will change only because of thermal diffusion. Hard rods, on the contrary, do develop a tendency to align and sediment in the direction parallel to the field. Their relevant mobility in this case is controlled by $\mu_\parallel$, which will be now a function of their separation. Their sedimentation behavior becomes involved because a pair of sedimenting rods will tend to rotate. Semiflexible filaments will combine their intrinsic tendency to rotate and develop a conformation perpendicular to the applied field with the additional deformation induced by the presence of the second chain. The details of their behavior depend on the geometrical distribution of the filaments [21].

Let us consider the particular case in which two collinear filaments lie perpendicular to an applied external field at the same initial height. The presence of a secondary filament will induce a rotation because the hydrodynamic stress is asymmetric. Such a translation–rotation coupling is also present in hard rods. For semiflexible filaments, this rotation will encompass additional filament deformation. The filaments will generically approach each other initially and will deform and rotate. Depending on its intensity, which will again be coupled to B, there is an additional deformation of the filament which will be asymmetric and, after a transient, may lead to filament repulsion. In Fig. 2a we show the $B - d$ diagram which displays if filaments approach or move away from each other as a function of their initial separation, d, for initially parallel chains. As B increases filaments move apart at smaller distances. Such a change of behavior cannot take place for hard rods, which will always approach each other. The only way in which approaching filaments can avoid collision is if their initial distance are such that they can rotate and become parallel to each other before colliding. In this case, the translation rotation coupling induces a rotation which leads subsequently to repulsion. Hence, rigid rods will perform a cyclic motion. Semiflexible filaments can exhibit such periodic orbits, but at large enough B then they will eventually move away from each other. This property is displayed in Fig. 2b, where we show the relative velocity of the two chains at time scales in which the fibers have displaced a distance of the order of their length. One can clearly see how, upon increasing B, filaments change their qualitative behavior and move away instead of attracting each other dynamically. At a rather large value of B a minimum in the repulsion velocity is observed, which signals the appearance of a third regime in which the repulsion becomes less sensitive to chain deformation.

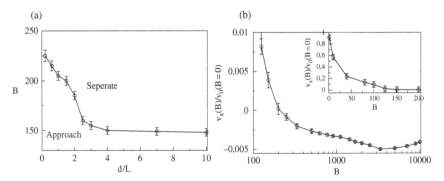

Fig. 2 (**a**) B-d diagram displaying the attraction or repulsion between two collinear filaments at long times. (**b**) Approaching velocities at $t = 20\tau_c$ when $d = L$; positive values imply attractive filaments. In the inseta the behavior at small B is shown in detail

The simplified geometries considered allow a quantitative analysis of the HI between filaments. The sedimentation velocity decays algebraically as the inverse of the filament distance and becomes quantitatively relevant for distances not larger than the typical filament size. The change of velocity due to the presence of a neighboring filament is bigger for the larger B, a cooperative effect due to the enhanced deformation by HI coupled to elastic deformation due to the inhomogeneous stress distribution acting along the filaments.

The filament deformation can be characterized by the maximum height difference between the filament's monomers, A. We have checked that $A(d) \sim d^{-3}$, providing further evidence of such a coupling between elastic deformation and HI.

5 Self-Propelling Particles

A complementary strategy to the one described in the previous sections is based on the dynamic resolution of the solvent. Rather than taking it implicitly into account through the propagator (e.g., the Oseen tensor in the model considered before), one is then interested in treating the solvent explicitly. One fruitful venue to accomplish such a goal is based on kinetic models. In particular, the lattice-Boltzmann approach (LB) has proven a flexible tool to study the dynamics of complex fluids [22].

In this approach one solves numerically a linearized Boltzmann equation which is discretized in space and time. Accordingly, the particle densities move on an underlying lattice and on each node the density of fluid particles is described in terms of the amount of particles, $f_j(\mathbf{r}_i, t)$, on node $\mathbf{r}$ moving with velocity $\mathbf{c}_j$. The set of allowed velocities, $\{\mathbf{c}_j\}$, joins a given node with a prescribed set of neighbors. The symmetry of the lattice and the minimum allowed set of velocities should conform to a minimum set of symmetry properties that ensure that the underlying anisotropy of the lattice does not affect the response of the system at the Navier–Stokes level [23]. The collision matrix determining the relaxation of $f_j(\mathbf{r}_i, t)$ is linearized with respect to the equilibrium distribution, $f_k^{eq}(\mathbf{r}_i, t)$,

$$f_j(\mathbf{r}_i + \mathbf{c}_j, t + 1) = f_j(\mathbf{r}_i, t) + \sum_k \Lambda_{jk}[f_k(\mathbf{r}_i, t) - f_k^{eq}(\mathbf{r}_i, t)] \qquad (8)$$

where the index k spans the velocity subspace, where Λ_{jk} defines the collision matrix which mixes the densities with different velocities at the corresponding node and ensures that both mass and momentum are conserved. This collision matrix is intimately related to the viscosities of the fluid. In the simplest model (referred to as exponential relaxation time (ERT)), $\Lambda_{jk} = \tau^{-1}\delta_{jk}$ is diagonal; the resulting LB is the lattice equivalent of BGK [24].

The simplicity of the dynamics has made LB a powerful tool. The connection with the collective dynamics of the fluid is done taking moments of these elementary densities, in the same way that hydrodynamic variables are obtained as moments of the one-particle distribution function in kinetic theory. For example, the local density $\rho(\mathbf{r}_i, t)$ and momentum $\rho\mathbf{v}(\mathbf{r}_i, t)$ are obtained from

$$\rho(\mathbf{r}_i,t) = \sum_k f_k(\mathbf{r}_i,t)$$

$$\rho(\mathbf{r}_i,t)\mathbf{v}(\mathbf{r}_i,t) = \sum_k \mathbf{c}_k f_k(\mathbf{r}_i,t) \tag{9}$$

where the index k runs over the subset of allowed velocities.

A suspended particle is described as a solid shell of a radius R, which is larger than the lattice spacing. The links which join exterior and interior nodes to the shell are bounced-back in the particle frame of reference at each time step. This additional rule enforces stick boundary conditions of the fluid at the solid surface. The change in velocity associated to the bounce back leads to a net force in the solvent. This force is precisely the opposite of the force that the fluid exerts on the solid. The torque the fluid exerts on the particle is computed in a similar way. Once this force is known, we can update the particle's velocity according to Newton's equations. This approach corresponds thus to a hybrid scheme in which the particles' dynamics is resolved individually.

In order to analyze microorganism motility, the spherical particles we model correspond to sizes of the order of the micrometer. The usual velocities of microoganisms lie in the range of the μm/s. Hence, inertia is irrelevant in their dynamics. Since they are supramicrometric, thermal fluctuations are also negligible. As a result, the particles can remain in suspension only due to their motility.

Since we are interested in the basic mechanisms controlling the collective dynamics of self-propelling particles, we introduce a simplified propulsion mechanism. On top of the standard bounce-back on the links, we subtract a constant amount of momentum with fixed magnitude Δp_0, uniformly distributed over all the fluid nodes connected to solid nodes and which lie within a cone of angle ψ_0 around a predefined direction of motion, as depicted in Fig. 3, where different propulsion mechanisms can be considered. The direction of motion determines the particle propulsion direction and moves rigidly with the propeller [25]. The subtracted momentum is added to the particle, ensuring momentum conservation. This particular

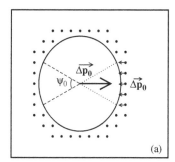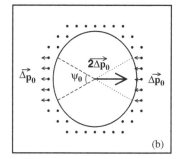

Fig. 3 Mechanism for particle propulsion. (**a**) Asymmetric driving. Momentum is subtracted from the nodes that lie within a cone of aperture ψ_0. The same amount is added with opposite sign to the particle. (**b**) Symmetric driving. The same mechanism as previously, but now momentum is subtracted from nodes from a cone with aperture ψ_0 symmetrically

choice of momentum transfer is inspired by internal peristaltic motion [26]. There is a great variety of propulsion mechanisms [27]. However, we will analyze general aspects of propulsion and cooperative motion which will not be sensitive to the details of how motility is produced.

This approach resolves the particles' shape, while other approaches enforce the fact that there is no net force generated by the propeller to generate its motion, but which neglect the finite size of the particles [28]. We can hence assess the role of short range interactions and excluded volume effects in the motion of propelling particles and their interplay with the HI.

A particle released from rest will reach a constant velocity, u_∞, after a time scale in which the fluid flow develops, $\tau_r \sim R^2/v$ for a particle of size R in a fluid of kinematic viscosity v. Consistent with this hydrodynamic origin, we have verified that the asymptotic velocity is reached algebraically, controlled by the long-time tail, $u(\tau) - u_\infty \sim \tau^{-d/2}$ $(d > 1)$, d being the system dimensionality. The asymptotic velocity arises from a balance between the driving force and the friction, $u_\infty = \Delta p_0/(12\pi\eta R)$, where $\eta = \rho v$ is the solvent viscosity [25]. The flow induced by the moving particle decays faster than $1/r$ because there is no net force exerted on the system. The decay is algebraic, but the specific form will depend on the particular type of driving considered. For the particular mechanism described above, the asymptotic flow field has dipolar symmetry.

If other particles are suspended in the fluid, they will interact through the fluid fluxes they generate due to their intrinsic motions. Such interactions will affect the dynamic properties of self-propelling suspensions. By analogy with colloidal suspensions, we can distinguish different dynamical regimes which characterize their collective dynamic properties.

6 Short-Time Dynamics

At times in which the induced fluid flow is developed, $t > \tau_r \sim R^2/v$, but still before the particles have displaced significantly, $t < \tau_m \sim R/u_\infty$, the presence of the other propellers will modify the effective velocity of the particles. In general, the additional drag induced by the flow generated by the neighbors leads to a decrease in the effective velocity, as displayed in Fig. 4. A randomly oriented distribution of self-propelling particles exhibits a very weak dependence on volume fraction. In electrophoresis, no dependence is expected for a homogeneous distribution of particles due to the absence of a net external applied force [29]. In these systems, however, the coupling between orientation and translation becomes very important, as we will discuss subsequently. One can see in Fig. 4 that when particles are oriented, the usual isotropy assumption disappears and a decrease in the mean velocity is observed. However, since the induced flow field decays faster than when a force is applied, this dependence is weaker than in sedimentation. The hydrodynamic coupling between propellers leads to fluid velocity distributions which deviate significantly from Gaussianity, as shown in Fig. 5. Although the fluid velocity

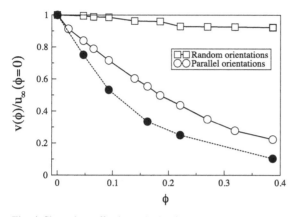

Fig. 4 Short-time effective velocity for a suspension of propellers moving at random and along a given direction. *Black symbols* correspond to the mean velocity for a passive sedimenting suspension

distribution function behaves as a Gaussian at velocities of order u_∞, a clear exponential is observed at small velocities while the asymptotic behavior is consistent with a generalized exponential. The behavior is more marked for the velocity distribution of the particles themselves, which behaves as power law for small velocity values. The crossover between these different regimes is determined by the characteristic velocity at which a particle moves, u_∞, where a small shoulder develops. Although deviations from Gaussian behavior are known in other non-equilibrium systems, e.g. driven granular fluids [30], in these suspensions they develop as a result of the internal mechanisms which generate motion and hydrodynamic coupling.

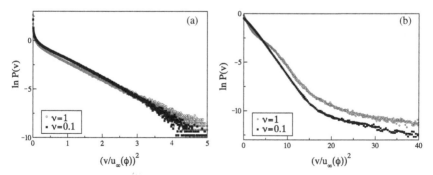

Fig. 5 Velocity distribution functions at volume fraction $\phi = 0.02$ in a system of $N = 1020$ self-propelling particles at two different solvent viscosities. Velocities are collected from independent initial conditions at $\tau/\tau_r \sim 2$. (**a**) Particle velocity distribution function; (**b**) fluid velocity distribution function

7 Long-Time Dynamics

On time scales of order τ_m and larger, the particles can displace over distances larger than their own size. In this time scale, structural relaxation processes take place. Some of these processes arise as a result of particles collision. However, the presence of the embedding fluid also gives rise to HI; as a result, particles feel each other and rearrange their relative positions through the drag forces they generate. We have observed that as particles approach each other they will modify their direction of motion and will tend to align. As a result, there is a tendency to move together only on the basis of their hydrodynamic coupling. Such alignment also leads to a transient acceleration; i.e. the particles will move faster than if they would be moving alone. This effect is different from the usual one observed for driven suspensions and arises as a result of the internal driving in these particles. Consistently, the angular velocity autocorrelation function decays typically in the collision time scale τ_m. Hence, propellers modify their direction of motion mostly through collisions. Although additional mechanisms that affect the propellers direction of motion, such as tumbling, play a relevant role in the decorrelation of particles' orientation, most of the features described in this contribution survive qualitatively if they are accounted for [25].

Due to subsequent collisions, both hydrodynamic and through direct particles' impact, the propelling particles will change their direction of motion and will diffuse at longer times. We can therefore identify an asymptotically long-time regime in which particles diffuse, characterized by the diffusive time scale $\tau_D \geq R^2/D$, where D is the particles' diffusion coefficient that arises through the described particle collisions. Since diffusion can only be achieved through successive collisions and relaxations to the local fluid flows, $\tau_D > \tau_m$, which implies that the propellers' Péclet number satisfies $Pe \equiv u_\infty R/D > 1$; hence for these systems convection is always a relevant mechanism. This feature is consistent with the long crossover regime observed in the mean square displacement (*msd*) of beads in a bacteria suspension [31]. In Fig. 6 we show the *msd* of a two-dimensional suspension of active particles at different volume fractions. We have marked the three characteristic time regimes, and the hierarchy of time scales, $\tau_r \ll \tau_m \ll \tau_D$ is clearly visible. For $t < \tau_m$ the *msd* increases quadratically with time, signaling the ballistic motion as a result of the fact that particles essentially move at a constant velocity which is proportional to u_∞ but decreases with volume fraction. When particles reach $t \sim \tau_m$ a crossover regime is observed. The precise value of the crossover region is also a function of the volume fraction. For small coverages the relevant distance is not the particle size but rather the interparticle distance, ℓ, which scales as $\ell \sim \phi^{-1/d}$ (d being the systems dimensionality). However, ℓ is a relevance distance for $\phi < 0.2$ for $d = 2$ (where $\ell/R = \sqrt{\pi/\phi}$) and it goes down to $\phi < 0.07$ for $d = 3$, where $\ell/R = (4\pi/(3\phi))^{1/3}$. During the crossover region induced by particle interaction one can clearly see an acceleration, in which the *msd* increases faster than ballistically. We attribute this acceleration to the cooperative hydrodynamic motion of groups of particles moving together; the hydrodynamic coupling increases particles' velocities. This complex interaction regime gives rise to a broad crossover regime where the collective motion

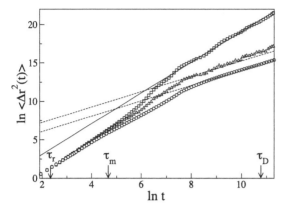

Fig. 6 Mean square displacement in a suspension of self-propelling particles for a two-dimensional system of 400 propellers, kinematic viscosity $\nu = 1$, and Reynolds number $Re = 0.25$. The *continuous straight line* corresponds to ballistic motion, $\langle \Delta r^2(t) \rangle \sim t^2$, while the *long-dashed ones* correspond to diffusion, $\langle \Delta r^2(t) \rangle \sim t$. *Circles* correspond to $\phi = 0.282$ whereas *squares* and *triangles* refer to $\phi = 0.1$ for different initial conditions. We also display the characteristic times: $\tau_r \sim R^2/\nu \sim 10$, $\tau_m \sim R/u_\infty \sim 100$, and $\tau_D \geq R^2/D \sim 5 \cdot 10^4$

is superdiffusive, in agreement with the observed crossover regime of the dynamics of passive beads in a suspension of motile microorganisms [31].

As particles come close together and increase their speed through HI, a translation-rotation hydrodynamic coupling also develops. As a result, propellers tend to move parallel to each other. This fact leads to an increase in particles' interaction time. While a small group of particles move together, they are likely to attract other moving propellers. As a result, we observe the appearance of clusters of particles on times larger than τ_m. The local volume fraction in these aggregates is quite large and not far from close packing. Hence, the smaller the mean volume fraction, the larger density contrast these suspensions will develop spontaneously. As a result, large transient aggregates are observed. In Fig. 7 we show a sequence of snapshots for an evolving suspension of mean volume fraction $\phi = 0.1$. We always start from uniform suspensions. One can see how transient clusters develop when particles start to interact with each other, and that very large clusters can form transiently. The faster than ballistic regime observed in the *msd* will happen during the formation of such structures. Since in our model there is no attraction between particles, these aggregates lack cohesiveness. As a result, when one of such aggregates collides with another one, it will typically dissemble partially. Moreover, the lack of cohesion implies that these aggregates are sensitive to stresses developed across them. In particular, the particles at the boundary are more sensitive to changes in local vorticity which will tend to decorrelate their orientation and may lead to particle disassembly. Hence, we cannot rule out that stresses impose an upper limit in the characteristic aggregate size. Nonetheless, numerically we have seen that aggregate disintegration is mostly controlled by cluster–cluster collisions.

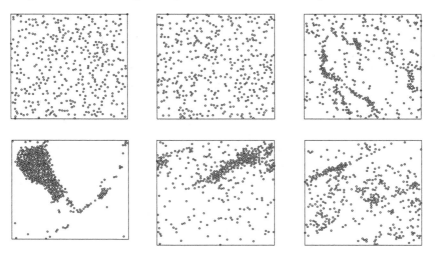

Fig. 7 Configurations of an initial homogeneous suspension of propellers at $\phi = 0.1$. The different times are, respectively, $t \sim 0.5\tau_m, t \sim \tau_m, t \sim 5\tau_m, t \sim 10\tau_m, t \sim \tau_D$, and $t \sim 2\tau_D$

Associated to this observation we have also noticed that for very small volume fractions clusters may span the system size, as shown in Fig. 8. In these situations the density contrast is so large that one cluster is able to attract most of the particles without competing with other clusters. Since our simulations have been run on systems sizes $L/R = 160$ containing up to 1000 particles, we cannot completely rule out the possibility that such percolating clusters are a finite-size effect. The existence of a characteristic size determined by the internal stresses would rule out these aggregates. However, in models of self-propelling particles in the absence of HI (in which case cluster formation is controlled by direct particle–particle attraction), it is known that non-equilibrium phase transitions exist leading to flocking [32]. In the

(a)   (b)

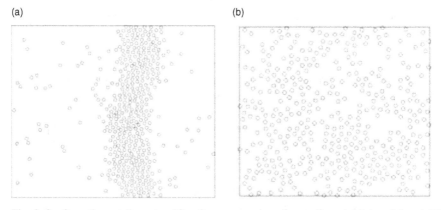

Fig. 8 Configurations at times $t \sim 10\tau_m$ for a suspension of propellers at (**a**) $\phi = 0.1$ and (**b**) $\phi = 0.4$

flocking regimes propellers move together and the system breaks the Galilean invariance leading to a net motion of the particles. In these flocking regimes spanning system clusters analogous to the one depicted in Fig. 8a are observed. Hence, the analogy with flocking transitions suggest that HI may induce analogous transitions. This issue requires a more detailed study to rule out finite size effects and the potential limitations of internally developed hydrodynamic stresses. The appearance of a flocking transition rules out the diffusion mechanism described earlier and controls asymptotically the suspension dynamics on time scales larger than τ_D. In fact, as shown in Fig. 6, for small volume fractions, if a size spanning cluster develops, the *msd* asymptotically behaves ballistically, consistently with the fact that Galilean invariance is broken and that the propellers suspension has acquired a net momentum. Hence, the existence of a characteristic time scale τ_D is intimately linked to the dynamics of hydrodynamically induced aggregate and their interactions. In our simulations we have always observed diffusion at long times for intermediate and large volume fractions. At large densities the tendency to form clusters is always present, but the smaller contrast with the background density prevents the development of larger clusters; a typical configuration is shown in Fig. 8b. A quantification of the transition between the different scenarios is certainly required. If additional interactions are present between particles, the relevance of these two different scenarios will vary. Obviously, the appearance of attractive interactions between mobile particles, i.e. through quorum sensing, will have a clear stabilization effect on the aggregates. Even if they are promoted initially through hydrodynamics, they can be stabilized and hence favor the development of large clusters which promote flocking. This mechanism will help the organisms to displace faster to new regions in the search of food. On the other hand, tumbling will have an opposed destabilizing effect and will prevent the appearance of such spanning clusters on time scales of the order of the inverse of the tumbling frequency.

In order to characterize more quantitatively the aggregates that propellers form spontaneously, we have computed the generalized radial distribution functions $g_n(r) \equiv \langle P_n(\cos\theta_{ij})\rangle$, with θ_{ij} being the relative angle between the direction of motion of the reference particle i and that of all particles j at a distance between r and $r+dr$, where P_n is the n-th degree Legendre polynomial. Figure 9a shows the lowest order radial distribution functions for an asymmetric driving and volume fraction $\phi = 0.1$. The tendency of moving particles to move close to each other is clearly seen in the value of the pair distribution function at contact, $g_0(r = 2R)$, which is much greater than its equilibrium counterpart. We display the equilibrium $g(r)$ for passive hard disks to show that the marked structure in the active suspension is not related with hard core interactions. In fact, the volume fraction is relatively small, so that hard core interactions play a negligible role developing structure. Superimposed to the large enhancement of the radial distribution at contact, the $g(r)$ also shows a decay on distances large compared to particle size. Such a decay is consistent with the presence of aggregates in the suspension of mesoscopic size. Transient aggregates are observed in other out-of-equilibrium systems, such as agitated granular fluids [33]. It is known that these aggregates can be understood as arising from large density fluctuations induced because of the lack of detailed balance.

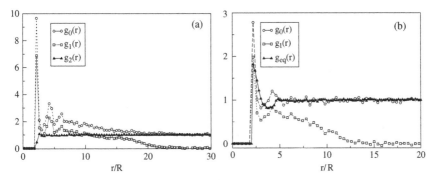

Fig. 9 (**a**) Radial distribution function, $g_0(r)$, of a self-propelling suspension for a two-dimensional system containing $N = 815$ particles, kinematic viscosity $v = 1$, and Reynolds numbers $Re = 0.25$ at times $\tau \geq 10\tau_m$ and volume fraction $\phi = 0.1$. (**b**) Radial distribution functions for an analogous system at $\phi = 0.1$, when the driving mechanism is symmetric around the direction of motion. Other generalized distribution functions, $g_n(r)$, are described in the text. *Thick black line* corresponds to the equilibrium radial distribution function

The connection between such aggregates in a variety of out of equilibrium systems remains an intriguing open question. The decay of $g_1(r)$ on the same length scale, indicates that there is a local tendency of propellers to develop a well-defined orientation, so that the aggregates have a common direction of motion. This implies that spherical moving particles are able to develop orientational order; their dynamics will display similarities with that of non-spherical propellers, where the tendency to develop local nematic order is shown to have profound implication in the rheological properties of these systems (see e.g. [34]).

We have also studied the collective behavior of symmetric propellers to assess the relevance of the specificity of the propulsion mechanism. In this case the momentum, Δp_0, is extracted uniformly from all nodes lying within the two cones defined by the angle ψ_0 and its supplementary, $\pi - \psi_0$, as shown in Fig. 3b. When compared to their asymmetric counterparts, the radial distribution function $g_0(r)$ shows a faster decay, implying that smaller clusters appear in this case. The contact value is still significantly larger than its equilibrium counterpart, although not as large as for the asymmetric driving. In this case $g_1(r)$ decays to zero on longer length scales than $g_0(r)$, which shows a tendency to local orientation of propellers, and hence a sensitivity toward nematization and which will affect its mechanical properties, beyond the structural constraint of forming well-defined aggregates.

Due to the clear tendency to the development of orientational ordering of propellers, even if spherical in shape, it is reasonable to compute the nematic order parameter as a measure of such ordering. At this point we will not distinguish between polar and apolar ordering, since we are only interested in clarifying the orientational ordering induced through hydrodynamic coupling between propellers. In Fig. 10 we display the time evolution of the nematic order parameter

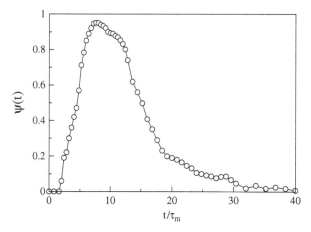

Fig. 10 Nematic order parameter, ψ, as a function of time for a suspension of propellers at $\phi = 0.1$ starting from a uniform, random configuration

$$\psi \equiv \frac{1}{N} \left\langle \sum_{j=1}^{N} \exp\left[i\theta_{ij}\right] \right\rangle, \tag{10}$$

where θ_{ij} is the angle between the directions of motion of particles i and j (remember that this direction may differ from the direction of motion of a particle). The plot shows a net increase in the order parameter on collision time scales, with values close to 1. As we enter in the diffusive time regime, the order parameter eventually decays slowly toward a rather small value.

Hence, the evidence collected shows how HI in suspensions of propelling particles make them approach and favor their cooperative motion because they favor particle alignment. Big clusters can be formed transiently, depending on the suspension volume fraction. In any case there is a high degree of particle orientation. Hence, it seems reasonable to assume that HI open a new way to induce flocking. Such a transition has different properties from the one described in the absence of coupling to the underlying fluid. In the latter case, the transition is promoted at high-volume fractions, when the direct interaction is stronger. On the contrary, hydrodynamics become more efficient at lower volume fractions, where higher density contrasts are possible and therefore where dynamically transient aggregates become more efficient to collect particles.

8 Conclusions

The use of mesoscopic models addresses quantitatively the dynamics of suspensions of biological interest in which the appropriate dynamic couplings with the solvent are properly captured. Since the molecular details are coarse grained, it is possible to achieve the time and length regimes that are relevant for the evolution of the suspended objects. Such methods are then suitable to the analysis of a variety of physical systems. In particular, in this chapter we have analyzed two complementary strategies to study the dynamics of inextensible semiflexible filaments and spherical propelling particles. The former constitute a good model for a variety of biofilaments (as, e.g., actin or microtubules) while the latter constitute a simplistic model of motile microorganisms.

We have shown how different approaches in which the solvent is either implicit or explicit are useful. For the case of filaments we have described how the interactions mediated through the solvent induce a strong cooperation in the dynamical response of these filaments. We have described how they rotate and show a rather rich response to applied field, an analysis that can be of interest in the analysis of results from ultracentrifuge or electrophoretic measurements. We have analyzed how semiflexibility gives rise to properties which are qualitatively different from those known for rigid rods.

The study of propelling suspensions has allowed to characterize their relevant dynamical regimes and their connection to well-known results of passive suspensions. We have shown the peculiarities of the hydrodynamic coupling at short time scales, and how the translation-rotation coupling leads to a larger dissipative interaction between propellers than expected from the known results from their passive counterparts. The coupling between translation and rotation at longer time scale leads to cluster formation and the development of mesostructures that can survive at long times, once the diffusive regime is achieved. Moreover, depending on the strength of their interaction and the volume fraction of the suspension, a transition to flocking-induced hydrodynamically is expected; although this particular issue requires more careful numerical studies.

This approach to the collective properties of suspensions is a promising tool to analyze the physics of a variety of systems with an obvious biological interest. The simplicity and local nature of the coupling between propellers and solvent allow one to envisage a number of relevant applications.

References

1. W. F. Pickard, Plant Cell Environ. **26**, 1 (2003).
2. D. Houtman, I. Pagonabarraga, C. P. Lowe, A. Esseling-Ozdoba, A. M. Emons and E. Eiser, Europhys. Lett. **78**, 18001 (2007).
3. M. L. Gardel, J. H. Shin, F. C. MacKintosh, L. Mahadevan, P. Matsudaira, D. A. Weitz, Science **304**, 1301 (2004).
4. D. Bray, *Cell Movements*, Garland, New York (2002).
5. J. Purcell, Am. J. Phys. **45**, 3 (1977).
6. A. Najafi and R. Golestanian, Phys. Rev. E **69**, 062901 (2004).
7. H. Wada and R. R. Netz, Europhys. Lett. **75**, 645 (2006).
8. B. Guirao and J. F. Joanny, Biophys. J. **92**, 1900 (2007).
9. A. Vilfan and F. Jülicher, Phys. Rev. Lett. **96**, 058102 (2006).
10. S. Gueron and K. Levit-Gurevich, Proc. Natl. Acad. Sci. USA, **96**, 12240 (1999).

11. B. Bassetti, M. C. Lagomarsino and P. Jona, Eur. Phys. J. B **15**, 483 (2000).
12. M. Doi and S. F. Edwards, *The Theory of Polymer Dynamics*, Oxford University Press, New York (1986).
13. C. P. Lowe, Fut. Gen. Comp. Syst. **17**, 853 (2001).
14. M. C. Lagomarsino, F. Capuani and C. P. Lowe, J. Theor. Biol. **224**, 215 (2003).
15. T. B. Liverpool, Phys. Rev. E **72**, 021805 (2005).
16. C. H, Wiggins, D. Riveline, A. Ott and R. Goldstein, Biophys. J. **74**, 1043 (1998).
17. M. P. Allen and D. J. Tildesley, *Computer Simulation of Liquids*, Oxford University Press, New York (1987).
18. X. Schlagberger and R. R. Netz, Phys. Rev. Lett. **98**, 128301 (2007).
19. R. Dreyfus, J. Baudry, M. L. Roper, M. Fermigier, H. A. Stone and J. Bibette, Nature **437**, 862 (2005).
20. M. Cosentino-Lagomarsino, I. Pagonabarraga and C. P. Lowe, Phys. Rev. Lett. **94**, 148104 (2005).
21. I. Llopis, M. C. Lagomarsino, I. Pagonabarraga and C. P. Lowe, Phys. Rev. E **76**, 061901 (2007).
22. S. Succi, *The Lattice Boltzmann Equation for Fluid Dynamics and Beyond*, Clarendon Press, Oxford (2001).
23. S. Wolfram, J. Stat. Phys. **45**, 471 (1986).
24. R. Verberg and A. J. C. Ladd, J. Stat. Phys. **104**, 1191 (2001).
25. I. Llopis and I. Pagonabarraga. Europhys. Lett. **75**, 999 (2006).
26. A. Ajdari and H. A. Stone, Phys. Fluids **11**, 1275 (1999).
27. S. Ramachandran, P. B. Sunil Kumar and I. Pagonabarraga, Eur. Phys. J. E **20**, 151 (2006).
28. J. P. Hernández-Ortiz, C. G. Stoltz and M. D. Graham, Phys. Rev. Lett. **95**, 204501 (2005).
29. J. L. Anderson, Ann. Rev. Fluid Mech. **21**, 61 (1989).
30. R. Brito and M. H. Ernst, J. Stat. Phys. **109**, 407 (2002).
31. X.-L. Wu and A. Libchaber, Phys. Rev. Lett. **84**, 3017 (2000).
32. J. Toner, Y. H. Tu and S. Ramaswamy, Ann. Phys. **318**, 170 (2005).
33. T. van Noije, M. H. Ernst, E. Trizac and I. Pagonabarraga, Phys. Rev. E **59**, 4326 (1999).
34. Y. Hatwalne, S. Ramaswamy, M. Rao and R. A. Simha, Phys. Rev. Lett. **92**, 118101 (2004).

Part III
Transport and Replication

A Thermodynamic Description
of Active Transport

S. Kjelstrup, J.M. Rubi and D. Bedeaux

Abstract We present a solution to problems that were raised in the 1960s: How can the vectorial ion flux couple to the scalar energy of the reaction of ATP to ADP and P, to give active transport of the ion; i.e. transport against its chemical potential? And, is it possible, on thermodynamic grounds to obtain non-linear flux force relations for this transport? Using non-equilibrium thermodynamics (NET) on the stochastic (mesoscopic) level, we explain how the second law of thermodynamics gives a basis for the description of active transport of Ca^{2+} by the Ca-ATPase. Coupling takes place at the surface, because the symmetry of the fluxes changes here. The theory gives the energy dissipated as heat during transport and reaction. Experiments are defined to determine coupling coefficients. We propose that the coefficients for coupling between chemical reaction, ion flux and heat flux are named thermogenesis coefficients. They are all probably significant. We discuss that the complete set of coefficients can explain slippage in molecular pumps as well as thermogenesis that is triggered by a temperature jump.

1 Introduction

A thermodynamic theory of transport may help describe cause–effect relations in biology. Ever since the work of Mitchell [1], when the proton motive force was given as the driving force for ATP synthesis, the driving forces for transport and the coupling of fluxes have been used in descriptions of energy conversion. This

S. Kjelstrup
Department of Chemistry, Norwegian University of Science and Technology, Trondheim, Norway,
signe.kjelstrup@chem.ntnu.no

J.M. Rubi
Departament de Física Fonamental, Universitat de Barcelona, Carrer Martí i Franqués 1, 08028-Barcelona, Spain, mrubi@ub.edu

D. Bedeaux
Department of Chemistry, Norwegian University of Science and Technology, Trondheim, Norway,
dick.bedeaux@chem.ntnu.no

Kjelstrup, S. et al.: *A Thermodynamic Description of Active Transport.* Lect. Notes Phys. **752**, 155–174 (2008)
DOI 10.1007/978-3-540-78765-5_7

makes non-equilibrium thermodynamics (NET), where these concepts are defined, a central theory [2, 3, 4, 5, 6, 7, 8, 9]. The classical version of NET is restricted to linear flux–force relations, however, so this theory has not been able to deal with energy conversion under all conditions. Observations show that flux–force relations are non-linear [5, 6, 7].

The aim of this work is to explain active transport; i.e. ion transport against a chemical potential difference by means of chemical energy, in terms of a recently proposed extension of classical NET. The extension is called mesoscopic non-equilibrium thermodynamics (MNET), because it addresses the level where molecular fluctuations take place (the meso-level) [10, 11, 12]. MNET introduces variables that are internal to the thermodynamic system and gives a thermodynamic basis for descriptions of phenomena in molecular motors and pumps. The central property of MNET is the same as for NET, the entropy production of the system. This property defines the fluxes and forces. Also MNET follows the systematic approach first given by Onsager [2, 3]. Local equilibrium is assumed at the appropriate level. Linear laws are written on this level and Onsager relations are used. Non-linearities will appear after integration from the meso- to the macroscopic level, as shown by Reguera et al. [10]. Active transport takes place at a membrane surface, because only a surface can account for coupling between a chemical reaction and transport of ions and heat [8].

The work is organized as follows. We present the problem from an overall perspective in Sect. 2. Active transport by the Ca-ATPase is used as an example throughout the analysis, drawing heavily on the experiments of de Meis and coworkers [13, 14, 15, 16, 17, 18]. In Sect. 3 follows a short summary of the use of NET over the years, an explanation of why this use leaves some problems unresolved, and what it takes to solve them. In the extension of NET in Sect. 4, which may solve the problems mentioned, we include as internal variable the well-known degree of reaction from reaction kinetics [19]. We finally define in Sect. 5 the experiments that determine the transport coefficients in active transport, and show that the complete set coefficients explains slippage in molecular pumps as well as thermogenesis triggered by a temperature jump.

2 Energy Conversion in the Ca-ATPase

In order to transport ions across membranes against their concentration gradient (i.e. active transport), biology uses chemical energy. This energy comes from hydrolysis of adenosine tri- phosphate (ATP), in short by

$$ATP + H_2O \rightleftarrows ADP + P \tag{1}$$

The products are adenosine diphosphate (ADP) and phosphate (P). The standard reaction Gibbs energy, ΔG^0, is around -33 kJ/mol [17]. The energy available is

$$\Delta G = \Delta G^0 + RT \ln \left[\frac{c_{ADP} \cdot c_P}{c_{ATP} \cdot a_w} \right] \tag{2}$$

Here R is the gas constant and T is the temperature. Concentrations of ATP, ADP and P are used, assuming that the ideal mixture approximation is good. All molar concentrations are divided by the concentration of the standard state (1 mol). The water activity $a_w = p/p^*$, where p is the vapour pressure of water above the solution, and p^* is the same over pure water.

Transport of ions means transport of charge. An electroneutral exchange of ions saves on the energy required to move ions against an electrical potential. It has been found that Ca^{2+} transport is countered by proton transport [14, 20]:

$$Ca^{2+} (i) + 2H^+(o) \rightleftarrows Ca^{2+}(o) + 2H^+(i) \tag{3}$$

where i stands for the outside (cytosol) and o for the inside (lumen) of the vesicle. We study the case where the energy of reaction (1) is used to accomplish (3), or vice versa, when (3) is used to drive (1). In the first case, the work accomplished is the chemical potential difference for exchange of the ions:

$$\Delta \mu_{Ca/2H} = RT \ln \frac{c_{Ca,o} \cdot c_{H,i}^2}{c_{Ca,i} \cdot c_{H,o}^2} \tag{4}$$

Charge numbers have been omitted for the ease of notation.

The chemical reaction takes place on the surface of the vesicle membrane in which the molecular pump, the Ca-ATPase, is embedded. A schematic picture of one Ca-ATPase with its membrane and cytosolic domains is shown in Fig. 1. A vesicle with Ca-ATPases forms spontaneously when the ATPase and phospholipids are mixed in a solution. The composition of the vesicle side o is controlled in the formation step. After forming, the vesicles can be harvested and dissolved in a test solution that defines the side i solution. In this manner one can investigate the operation of the molecular pump under several conditions. Of interest is in particular the reaction rate, r, and the ion flux, J_{Ca}. A positive measurable heat flux from the Ca-ATPase to the solution phase has been found when the reaction rate is positive [16].

The area of the membrane surface is generally not known. It is therefore assumed that the area is proportional to the amount of Ca-ATPase that is reconstituted in the membrane. Fluxes as well as the reaction rate are therefore measured per milligram of protein. The ratio of the ion flux from the Ca-ATPase into the vesicle, and the reaction rate, can be determined:

$$\frac{J_{Ca}^o}{r} = n \tag{5}$$

When n is an integer, equal to the number of binding sites for the ion on the pump, the pump is said to be stoichiometric.

Fig. 1 A schematic picture of
active transport by the
Ca-ATPase

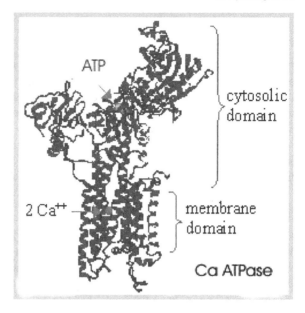

ATP

cytosolic
domain

$2\,Ca^{++}$

membrane
domain

Ca ATPase

In our example the number of binding sites for Ca^{2+} is two, see Fig. 1. A non-stoichiometric pump, for which n is smaller than the number of binding sites, is said to *slip*. The value $n = 2$ has been measured only for small values of $\Delta\mu_{Ca/2H}$. The further n is from 2, the higher is the degree of slip in the pump, the smaller amount of work is done, and the larger is the heat that is produced. This can be understood from the following.

The lost work (the dissipated energy) plus the actual work is the maximum work available in the system, i.e. a constant. For an isothermal, isobaric process, the maximum work available W_{ideal} is the reaction Gibbs energy [9]. In non-isothermal systems, there are additional contributions to W_{ideal}. No system is able to use all of the available energy to do work, so $n\Delta\mu_{Ca/2H} < -\Delta G$. According to the second law of thermodynamics, a fraction is always dissipated as heat. The entropy production in the system measures this heat production. The product $W_{lost} = T_0\,(dS_{irr}/dt)\,\Delta t$ is the lost work, or the energy dissipated as heat [9] during the time Δt. Here T_0 is the temperature of the surroundings (in K), and dS_{irr}/dt is the entropy production in (J/s K m^3). We have

$$W_{ideal} = -\Delta G = W + T_0 \left(\frac{dS_{irr}}{dt} \right) \Delta t \tag{6}$$

where W is the real work performed (in J) during Δt. The ratio between work and heat production can vary largely [16], but is bounded by W_{ideal}. The larger the fluxes and forces are, the larger is the entropy production, see Sect. 4.1. The terms in (6) are determined as follows. The control volume V has a known concentration of Ca-ATPase, i.e. pumps (in mg/m^3). The pump is observed during stationary state operation in a time interval Δt. The absolute amount of ATP consumed in this interval is measured (in mol/mg). The central term in (6) can be determined as the difference in Gibbs energy (in J/m^3) between the product state and the reactant state. The Ca-flux can also be determined in the experiment, leading to a similar calculation of the work done, W, from (4).

The entropy production is a product sum of all independent fluxes and their conjugate forces. Consider as an example, the electric current density (in C/s m^2) as a flux. The conjugate driving force to this flux is minus the electric field divided by the temperature (in V/m K). Their product times the temperature of the surroundings gives the Joule heat in the volume element of the conductor times the factor T_0/T. The entropy production is like a generalised friction due to transport through the volume element. If the friction occurs at a temperature higher than the surroundings, some of the heat produced can be used to give work with respect to the surroundings.

Biological systems are mostly regarded as isothermal, i.e. without temperature gradients, and no distinction is made between T and T_0. A heat flux has so far not been considered a central variable in active transport, in spite of clear documentations of heat effects [13, 14, 16].

In the analysis of a spherical vesicle with active transport of Ca^{2+} across the vesicle membrane, take for simplicity $T_0 = T$. With leaky, isothermal vesicles, no work is done on the vesicle ($W = 0$ in (6)) and all available energy $(-\Delta G)$ is dissipated as heat in the system plus surroundings. In this case the entropy production is the product of the reaction rate and the reaction Gibbs energy over the temperature. The energy dissipated per unit of time at this temperature is then

$$T \frac{dS_{irr}}{dt} = -r\Delta G \tag{7}$$

The dimension of the products on both sides are here J/s mg.

The molecular pump can also be operated in a reverse mode. This means that the osmotic energy can be used to synthesize ATP. De Meis and coworkers [13, 14] also showed that a pH-jump and a change in water activity can drive synthesis of ATP. The effect of these variables follows from (2) and (4) above. Much more difficult is to explain a similar observed effect on the ATP synthesis by a temperature jump [13, 14].

According to NET [8, 9, 22], it is fully possible to have a heat flux, also near isothermal conditions, in this system, because the heat flux can be coupled to the reaction rate and the calcium flux at a membrane surface. Such a heat flux shall be included as a variable in our use of this theory. We shall see that it may give a quantitative basis for observations on thermal effects, in this particular case. We believe that the phenomenon is quite general, however.

3 Towards a Thermodynamic Transport Theory

3.1 Short Background

A non-equilibrium thermodynamic (NET) description of active transport was first proposed by Katchalsky and Curran [4]. Assuming that active transport involves a chemical reaction taking place *within* a membrane, they gave the energy dissipation in the system in terms of forces and fluxes for the whole membrane:

$$T\frac{dS_{irr}}{dt} = -IE - r\Delta G \tag{8}$$

The net ion flux was I and E was the electromotive force[1]. No heat flux was included, as the thermal force was considered to be zero. Linear force–flux equations were written for the whole membrane. The ion flux and the reaction rate were both treated as scalars. Coupling is then formally possible, but has no local stochastic basis.

More in-depth understanding was offered by Caplan and Essig [5] and later Westerhoff and van Dam [6]. Caplan and Essig introduced so-called proper pathways. These paths describe how a system relaxes when a stationary state is perturbed, also far from global equilibrium. For instance, there may be a quasi-linear region around the inflection point in a plot of r versus ΔG. The equations for active transport were written as

$$r = -L_{rr}\Delta G - L_{rd}\Delta\mu_{Ca}$$
$$J_{Ca} = -L_{dr}\Delta G - L_{dd}\Delta\mu_{Ca} \tag{9}$$

As driving force for ion transport was used the chemical or the electrochemical potential difference. Linear relations have since then been used in non-equilibrium thermodynamics to describe fluxes and forces near saturation of ion binding sites, also far from equilibrium, see e.g. Waldeck et al. [23]. A stoichiometric pump, where (5) applies, has a singular matrix of coefficients

$$L_{dd} = 2L_{dr}$$
$$L_{rd} = 2L_{rr} \tag{10}$$

Relations like (9) can be derived as limiting cases from the more general Hill diagrams [5, 30]. These diagrams describe stationary state transport phenomena by combinations of unidirectional pathways. Each unidirectional path is described by its transition probability, and the equation for the net transport is given as combination(s) of two or more unidirectional fluxes. Non-linear flux–force relations were easily found and explained by this theory. While these formulations are correct and important, it is fair to say that they provide the original problem of the coupling of

---

[1] The authors used the affinity $A = -\Delta G$, and did not specify the temperature T.

scalars and vectors with a kinetic, but not a thermodynamic basis. This has motivated us [11, 12] to seek for a different way to predict non-linear behaviour, as an alternative to the work of Hill [30], using the extension of classical non-equilibrium thermodynamics, MNET.

3.2 A Stochastic Basis and an Internal Variable

In search for a thermodynamic basis for active transport, we observe first that the underlying events of the energy transfer are stochastic; they occur on the molecular level. A description must therefore address the time-scale and the variables on this level. It is therefore interesting that the conformational changes in the Ca-ATPase during pump operation take place very slowly, on a milli-second time-scale [20]. At least five different states are involved, as has been shown by X-ray crystallography [21]. Intermediate states must be included in the definition of the time-scale for the coupled transports. It does not give meaning to use a two-state description of the chemical reaction [11, 12] with the states of the products and the reactants only.

A natural variable for the path given by the conformational changes is the degree of reaction, γ. Kjelstrup et al. [11, 12] chose this variable as a measure for the scale where fluctuations take place. The degree of reaction, γ, is a standard variable in reaction kinetics [19]. The path in question is illustrated in Fig. 2. The lower part of the figure shows the activation energy barrier, which is very large in the present case (85 kJ/mol) [20]. There is an equilibrium distribution for the reacting mixture, i.e. the composition of reactant and products over the barrier, as indicated in the upper part of the figure. The amount of the mixture is minimum at the coordinate of the transition barrier, γ_{tr}. Such a distribution can also be defined away from equilibrium. By introducing the degree of reaction as an internal variable in the thermodynamic description, see de Groot and Mazur for this type of problems [22], it was possible to give the stochastic nature of the process a proper scale [11, 12].

3.3 Coupling of Scalar Components at the Surface

In order to have coupling between reaction, diffusion and heat transport, all phenomena must take place on the same scale. This scale was discussed above. But coupling of scalars and vectors are not allowed in homogeneous systems. A certain anisotropy or symmetry breaking is needed in order to have coupling of scalar and vectorial processes (the so-called Curie-principle, see e.g. de Groot and Mazur [22]). There is a possibility for coupling between chemical reaction and vectorial phenomena at surfaces, because the surface has no symmetries in the direction normal to the surface [8]. The normal component of the three-dimensional vector has

Fig. 2 The activation energy
barrier for the chemical
reaction

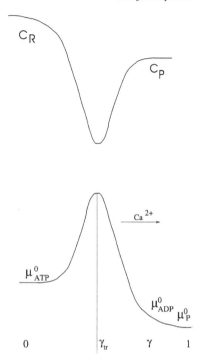

a symmetry of a scalar along the surface. This property is sufficient to give a formal basis for coupling between chemical, electrical, thermal or diffusional forces in the surface. In the present case, the chemical reaction (1) takes place in the surface, while the fluxes that couple to r are normal components into or across the surface.

It follows that the thermodynamic system of interest here is a surface element, and this must first be defined. We follow Gibbs who defined surfaces as two-dimensional systems. The surface is treated as an autonomous, two-dimensional system with excess variables. Thermodynamic variables for the surface like the concentrations were given as moles per surface area, in excess of the extrapolated bulk concentrations. Density variations across the surface are for this purpose integrated out over the surface thickness and thereafter allocated as excess variables of the surface element.

The thermodynamic surface that we are speaking of in Fig. 1 must contain the region that includes the binding sites for ATP as well as for Ca^{2+}. By integrating over this region, we shall obtain the excess variables discussed above, as for instance the concentration of the protein. Coupling must involve interaction between the two binding sites, and takes place, when there is a large degree of correlation of the fluctuations of the relevant variables in the surface.

3.4 Local Equilibrium Along the Path of the Chemical Reaction

The assumption of local equilibrium is always implicitly used when thermodynamic equations are written in NET. In order to have thermodynamic properties defined, a system must have enough particles for a statistical analysis. The surface element containing the Ca-ATPase is small, so the assumption is questionable. Not only is the surface small, one vesicle contains only a few Ca-ATPases. The number of vesicles in solution is usually large, however. Hill [29, 30] defined thermodynamic equations for small systems. By considering replicas of a system, he constructed an ensemble for statistical analysis. Here the single system in Hill's sense is one Ca-ATPase, and the replicas are all Ca-ATPases, distributed over all vesicles. The average behaviour of one Ca-ATPase can then be defined. We adopt this picture of Hill, and write thermodynamic relations for excess thermodynamic properties of the ensemble of vesicle surface elements in equilibrium.

But can we use these equations also when the surface is exposed to large gradients? The biological membrane is thin (6–8 nm), so gradients across the surface can be enormous. Evidence from non-equilibrium molecular dynamics simulations now give some support for a positive answer [27, 28].

The assumption of local equilibrium along the path of reaction may also be questioned. We shall argue that the many isolated states of the Ca-ATPase [21] and the slow and reversible reaction make the assumption of local equilibrium probable, at any position along the path of the chemical reaction in Fig. 2.

3.5 Describing Active Transport: Criteria

Let us summarize the discussion by stating again the premises for giving a thermo-dynamic basis to active transport. We require the following:

- A surface element where coupling takes place between reaction and ion transport. This gives a formal basis for coupling.
- A variable that deals with the progress of chemical reaction over its many stages. This gives a stochastic basis for coupling.
- Local equilibrium along the path of the chemical reaction. This enables us to use NET on a mesoscopic level.
- The heat flux as a variable, in spite of the system being (nearly) isothermal. This opens up a possibility to describe thermal effects associated with active transport.

4 Mesoscopic Non-equilibrium Thermodynamics

We can now proceed (in Sect. 4.1) to find the entropy production of a surface element with a chemical reaction, an ion flux, and a heat flux. This will next determine (in Sect. 4.2) the flux equations of the system.

4.1 The Entropy Production in the Surface

Entropy production occurs in the solution (i), in the surface (s) and in the vesicle membrane. We neglect the entropy production by ion transport in the membrane, because this process is fast. Equilibrium for binding/debinding of ATP, ADP and P to the reaction site, and for binding of two Ca^{2+} to the binding site for ions, means that we also can neglect entropy production in the solution. All of the entropy production during active transport is then assumed to take place in the surface. The Gibbs equation for the entropy density, s^s, of the surface is

$$ds^s = \frac{1}{T^s} du^s - \frac{\mu^s_{ATP}}{T^s} dc^s_{ATP} - \frac{\mu^s_{ADP}}{T^s} dc^s_{ADP} - \frac{\mu^s_P}{T^s} dc^s_P - \frac{\mu^s_{Ca/H}}{T^s} dc^s_{Ca} \qquad (11)$$

where $\mu^s_{Ca/H} = \mu^s_{Ca} - 2\mu^s_H$, cf. (3) $(2dc^s_{Ca} = -dc^s_H)$. We assume that the total entropy of the membrane together with the solution is unaltered if the various components are moved from the solution to the surface and back, without moving internal energy (partial equilibrium for adsorption/desorption; cf. de Groot and Mazur [22], Chap. 15). The assumption of isentropy (equilibrium in this part of the open system) implies that all the Planck potentials (μ/T) for the surface can be replaced by their solution counterpart. On the other side of the surface we assume that internal energy can be moved, without moving ions, without altering the entropy. This implies that the surface temperature is equal to the temperature inside the vesicle, giving $T^s = T^o$. These assumptions are introduced in (11) and the result is applied to any value of the degree of reaction, γ. For a given value of γ, we then have [11]

$$ds^s(\gamma) = \frac{1}{T^o} du^s(\gamma) - \frac{\mu^i_{ATP}}{T^i} dc^s_{ATP}(\gamma) - \frac{\mu^i_{ADP}}{T^i} dc^s_{ADP}(\gamma) - \frac{\mu^i_P}{T^i} dc^s_P(\gamma) - \frac{\mu^i_{Ca/H}}{T^i} dc^s_{Ca}(\gamma)$$
$$(12)$$

where $s^s(\gamma)$ and $u^s(\gamma)$ are the excess entropy and internal energy of the Ca-ATPases in state γ. Furthermore, $c^s_i(\gamma)$ is the probability that the excess of component (i) is adsorbed in an Ca-ATPase in the state γ times the total adsorption of this component. By integrating $ds^s(\gamma)$ over γ from 0 to 1, we recover ds^s, and similar for the other variables. The mass and energy balances reflect that the process is stochastic:

$$\frac{dc_r(\gamma)}{dt} = -\frac{dr(\gamma)}{d\gamma} \qquad \frac{dc^s_{Ca}(\gamma)}{dt} = J^i_{Ca}(\gamma) - J^o_{Ca}(\gamma) \qquad (13)$$

$$\frac{du^s(\gamma)}{dt} = J^i_q(\gamma) - J^o_q(\gamma) \qquad (14)$$

In these equations $r(\gamma)$ is a reaction flux along the γ-coordinate. Furthermore, $J^v_{Ca}(\gamma)$ and $J^v_q(\gamma)$ are fluxes that directly exchange calcium ions and heat with the v phase from the Ca-ATPases in the state γ; they do not flow along the γ-coordinate. Escape of Ca^{2+} to the o-side, with $r > 0$, means that the pump is slipping.

The entropy production of the surface as a function of the internal variable or coordinate becomes:

$$\sigma^s = -r(\gamma)\frac{1}{T^i}\frac{\partial G(\gamma)}{\partial \gamma} - J^o_{Ca}(\gamma)\left(\frac{\mu^o_{Ca/H}}{T^o} - \frac{\mu^i_{Ca/H}}{T^i}\right) - J^i_q(\gamma)\left(\frac{1}{T^o} - \frac{1}{T^i}\right) \quad (15)$$

Equation (15) explains that entropy is produced along the γ-coordinate of the chemical reaction, by the ion transport and by heat transport. Due to the existence of the potential barrier along the γ-coordinate, see Fig. 2, a quasi stationary state is set up during the reaction in which $dr(\gamma)/d(\gamma)$ is zero and equal to r. We can then integrate (15) under stationary state conditions and obtain the entropy production in the surface for the conditions chosen:

$$\sigma^s = -r\frac{1}{T^i}\Delta G - J^o_{Ca}\left(\frac{\mu^o_{Ca/H}}{T^o} - \frac{\mu^i_{Ca/H}}{T^i}\right) - J^i_q\left(\frac{1}{T^o} - \frac{1}{T^i}\right) \quad (16)$$

This applies to active transport of Ca^{2+} by the Ca-ATPase, with the assumptions stated. The fluxes J^o_{Ca} and J^i_q are the integrals of $J^o_{Ca}(\gamma)$ and $J^i_q(\gamma)$ over γ. The stationary entropy production is also given by the entropy balance over the surface element:

$$\sigma^s = \frac{J'^o_q}{T^o} - \frac{J'^i_q}{T^i} + r\Delta S + J_{Ca}\Delta S_{Ca/H} \quad (17)$$

where J'_q denotes a measurable heat flux. This equation explains that entropy produced in the surface, σ^s, comes from the reaction in the surface ($r\Delta S_i$), the exchange of Ca^{2+} with 2 H^+, and measurable heat fluxes over their respective temperature. The measurable heat flux on the i-side can be found from the energy balance.

$$J^i_q = J'^i_q + r\Delta H - J_{Ca}H^i_{Ca/H} \approx J'^i_q + r\Delta H \quad (18)$$

It is equal to the total heat flux (or the energy flux) minus the enthalpy change by the reaction and the enthalpy change by the exchange of Ca^{2+} with 2 H^+. In agreement with the ideal mixture approximation used in (1), we neglected the enthalpy for exchange of Ca^{2+} with $2H^+$. When the system in question has equilibrium in the chemical reaction, the expression simplifies further. We then have $r\Delta H = rT\Delta S \approx 0$.

In isothermal calorimetry, one measures the heat production at constant temperature. The measurement gives the integral of J'^i_q over time. The sensible heat flux, J'^o_q, directed to the vesicle inside will eventually leak back, not only through the ion channel, but through all of the surface area available on the spherical vesicle. This will disturb the experiment. One must account for the fact that the transport across a real vesicle is *not one-dimensional*.

4.2 Flux Equations

The entropy production determines the fluxes and forces of the system. According to standard non-equilibrium thermodynamics, each flux is written as a linear combination of each force. Now this must be done in γ-space (not done here, for details see [12]). The coefficient matrix in γ-space is symmetric according to Onsager.

The fluxes are next integrated over the γ-coordinate, using Kramers' rule [25][2]. The integrated version of the flux equations gives the expressions that can be used to reduce experimental data:

$$r = -D_{rr}(1 - \exp\frac{-\Delta G}{RT^i}) - D_{rd}\left(\frac{\mu^o_{Ca/H}}{RT^o} - \frac{\mu^i_{Ca/H}}{RT^i}\right) - D_{rq}\left(1 - \frac{T^i}{T^o}\right) \quad (19a)$$

$$J_{Ca} = -D_{dr}(1 - \exp\frac{-\Delta G}{RT^i}) - D_{dd}\left(\frac{\mu^o_{Ca/H}}{RT^o} - \frac{\mu^i_{Ca/H}}{RT^i}\right) - D_{dq}\left(1 - \frac{T^i}{T^o}\right) \quad (19b)$$

$$J^i_q = -D_{qr}(1 - \exp\frac{-\Delta G}{RT^i}) - D_{qd}\left(\frac{\mu^o_{Ca/H}}{RT^o} - \frac{\mu^i_{Ca/H}}{RT^i}\right) - D_{qq}\left(1 - \frac{T^i}{T^o}\right) \quad (19c)$$

The forces were made dimensionless, for the sake of simplicity. The coefficients have, then, the same dimension as the flux they describe. The first force is the *chemical driving force* for the activated reaction, the second is the *osmotic driving force* for the ion flux, and the third force is the *thermal driving force*. The non-linearity appears as a consequence of the activated process, cf. Fig. 2.

The coefficient matrix is no longer symmetric. The D_{ij}-coefficients are still related because Onsager relations are valid on the γ−scale. We have now

$$D_{dr} = AD_{rd} \quad (20)$$

and

$$D_{qr} = ART^i D_{rq} \quad (21)$$

where A is a constant (see [12] for an explicit expression) of the experiment, and

$$D_{dq} = D_{qd}/RT^i \quad (22)$$

These relations reduce the number of unknown coefficients to six. The coefficients are functions of state variables like the temperature and concentration, but are not functions of the forces. The non-linearity is due to the chemical force only, and it is *not* a sign of being far from equilibrium. On the contrary, we have assumed that there is local equilibrium everywhere along the γ-coordinate in the surface.

It is interesting that the fluxes are linear in two of the forces, and non-linear only in the chemical force. This may be one reason why linear flux force relations have been found more often than expected [23]. When $\Delta G << RT$ the chemical driving force simplifies, and the fluxes become linear in all forces.

---

[2] Constant $J_H = -2J_{Ca}$ and r, and a constant mobility along the path.

4.2.1 Active Transport

With reference to (19), active transport occurs because the coefficients D_{rd} and D_{dr} are non-zero. The larger the coefficients are, the better is the energy transfer between the chemical driving force and the osmotic driving force. The pump is reversible, thanks to these coefficients.

The chemical driving force and the reaction rate are positive during active transport, while the osmotic driving force is negative, meaning that D_{rd} and D_{dr} are positive. A positive osmotic force will then give rise to a negative reaction rate (i.e. ATP synthesis) even when $\Delta G > 0$.

A stoichiometric pump, with $J_{Ca}/r = 2$, has in addition to (20)–(22), also

$$D_{dd} = 2D_{rd} \qquad D_{dr} = 2D_{rr} \qquad D_{dq} = 2D_{rq} \qquad (23)$$

These relations define the stoichiometric pump for a wider range of states than before. They can be regarded as extensions of (10), and give an upper limit of the coupling coefficients.

$$D_{dr} \leq 2D_{rr} \qquad D_{rd} \leq D_{dd}/2 \qquad (24)$$

When the equal signs apply, and the system is isothermal, one may further verify that the ratio between the total heat flux and the reaction rate is also constant.

$$\left(\frac{J_q^i}{r}\right)_{\Delta T=0} = m \qquad (25)$$

This lower limit for the total heat flux was not given earlier. The entropy production becomes

$$\sigma^s = -r\frac{1}{T^i}\left[\Delta G - 2\left(\mu_{Ca/H}^o - \mu_{Ca/H}^i\right)\right] \qquad (26)$$

A stoichiometric pump has a smaller entropy production per ATP hydrolysed than a non-stoichiometric one, see (16). According to Ross and Mazur [24], the entropy production is a product sum of fluxes and forces also when the chemical reaction is in its non-linear regime.

4.2.2 Static Head and Level Flow

The linear flux–force relations of (9) have been used to define the condition of static head and level flow.

The static head condition can be compared to the state of an electrochemical cell with negligible current drawn, with a balance of chemical and electromotive forces. Here we have instead a balance of the chemical and the osmotic force and a zero flux of Ca^{2+} (near zero entropy production). The condition for static head from (19) is

$$0 = -D_{dr}\left(1 - \exp\frac{-\Delta G}{RT^i}\right)_{sh} - D_{dd}\left(\frac{\mu^o_{Ca/H}}{RT^o} - \frac{\mu^i_{Ca/H}}{RT^i}\right)_{sh} - D_{dq}\left(1 - \frac{T^i}{T^o}\right)_{sh} \quad (27)$$

where sh stands for static head. The relation applies for small as well as large chemical forces.

The level flow condition, on the other hand, can be compared to a short-circuited electrochemical cell. This translated to a case with zero difference in chemical potential of Ca^{2+}. In the presence of level flow, when $\left(\frac{\mu^o_{Ca/H}}{T^o} - \frac{\mu^i_{Ca/H}}{T^i}\right) = 0$, the ion flux is driven by the chemical and thermal forces,

$$(J_{Ca})_{lf} = -D_{dr}(1 - \exp\frac{\Delta G}{RT^i}) - D_{dq}\left(1 - \frac{T^i}{T^o}\right) \quad (28)$$

where lf means level flow. The equation applies for large as well as small chemical forces. A similar equation can be written for the reaction rate.

The form of the entropy production will be the same in both cases:

$$\sigma^s = -r\frac{\Delta G}{T^i} - J^i_q\left(\frac{1}{T^o} - \frac{1}{T^i}\right) \quad (29)$$

But its value will vary between the cases. If we consider the first term only, we have for static head

$$\sigma^s_{sh} = -\frac{\Delta G}{T^i}\left[D_{rr} - \frac{D_{rd}D_{dr}}{D_{dd}}\right]\left(1 - \exp\frac{-\Delta G}{RT^i}\right) \quad (30)$$

while the entropy production at level flow is higher:

$$\sigma^s_{lf} = -\frac{\Delta G}{T^i}D_{rr}\left(1 - \exp\frac{-\Delta G}{RT^i}\right) \quad (31)$$

These equations apply when the Ca-ATPase is reconstituted in tight vesicles. When a leak pathway is opened up for Ca^{2+}, we may also encounter similar conditions (sh or lf). The transport coefficients change, however.

4.2.3 A Slipping Pump

A stoichiometric pump was defined above, as characterised by coefficients which obey (25) and (23). Away from these conditions, the coupling coefficients are smaller, cf. (24). The coupling is not so good, when there is a higher probability for the ion to escape its binding site in the direction of its downhill gradient, rather than being transported uphill. Less work is also done when the coupling coefficient

D_{rd} is smaller; there is more friction in the pump and the entropy production becomes larger.

The flux–force relations are non-linear for both the stochiometric and the slipping pumps. We cannot deduce that this property alone is the origin of a slipping pump, as has been done in the past. But, it is probable that slip is more likely when one of the driving forces is large. The fast process will try to equilibrate on a smaller time-scale than the other. The slow process will then loose energy into the fast, and the coupling will be less complete.

4.2.4 Thermogenesis

The set of equations show that a total heat flux is always present, even when the thermal force is zero. The only requirement is that D_{qd} and D_{qr} are non-zero. Active transport, as we know it, is then always associated with a heat flux to the solution. This total heat flux can to some degree be reversed. We see from (19) that it depends on the sign of the forces, like the reaction rate and the ion flux do. That heat effects are associated with charge transport, is well known from electronic conductors (the Peltier effect). Reversible heat effects are interpreted by entropy changes in electrochemical cells [31].

The magnitude of the total heat flux into the solution, $-J_q^i$, depends on several variables: the forces as well as the coefficients. As we discussed above the coupling coefficients have an upper limit, when the pump is stoichiometric. As the pump starts to slip, the coupling coefficients become smaller. Less work is done, and heat production increases.

Several explanations have been given for thermogenesis in the literature. It has been associated with the normal heat production that accompanies physical activity. It has also been linked to changing dietary conditions. Going back to the relation between available work, work and lost work, (6), there is always a non-zero entropy production related to the resting state of the organism, or the system. The entropy production of the resting state will change once any of the driving forces changes: the chemical force, the osmotic force, or the thermal force. This changes immediately the ratio of work done to heat produced. These arguments address only physical properties of the ATPase, not any extra biological regulation.

Thermogenesis can be then defined as an increase in the entropy production above the resting state level, caused by a change in the thermal force. Such a change in the force can be realised, for instance, by a temperature drop across the vesicle membrane.

Thermogenesis is possible with non-zero coupling coefficients D_{rq}, D_{qr} and D_{dq}, D_{qd} in equation (19). The first two coefficients explain that a temperature drop triggers the ATPase reaction (1). This leads to heat production in the solution via D_{qr}.

5 Experimental Determination of Transport Coefficients

One advantage of the theory of NET is the possibility it gives to define experiments. This shall be done next. The transport coefficients are not functions of the driving forces, but they may be functions of the state variables. Before more data are available, we shall assume that they are constant for reasonable changes in the state variables. A coefficient, determined for one set of conditions, can then be used also for somewhat different conditions.

5.1 The Main Coefficients

There are three main coefficients in the matrix for transport coefficients; the diagonal coefficients; These must all be determined independently. The coefficient D_{rr} describes how the rate depends on the chemical driving force. It can be measured in the absence of a temperature difference and a osmotic force.

$$D_{rr} = - \left[\frac{r}{1 - \exp(-\Delta G_i / RT)} \right]_{\Delta \mu_{Ca/H} = 0, \Delta T = 0} \tag{32}$$

The reaction rate as well as the reaction Gibbs energy can be determined with good accuracy. The measurement conditions are probably obeyed with leaky vesicles [14].

The interdiffusion coefficient of Ca^{2+} with $2H^+$ is determined at zero chemical and thermal force:

$$D_{dd} = - \left[\frac{J_{Ca}^o}{\dfrac{\mu_{Ca/H}^o}{RT^o} - \dfrac{\mu_{Ca/H}^i}{RT^i}} \right]_{\Delta G = 0, \, \Delta T = 0} \tag{33}$$

The definition says that the ion flux should be determined at chemical equilibrium (with a small reaction rate) and for isothermal conditions. This experiment is difficult without invoking a sizable reaction rate. One must thus correct for this rate. It is however likely that the self-diffusion coefficient for Ca^{2+} gives a good estimate for the order of magnitude of the exchange rate of the two ions. Self-diffusion can be measured with good accuracy using radioactive isotopes.

The thermal conductivity coefficient D_{qq} is defined for an inactive pump:

$$D_{qq} = - \left[\frac{J_q^i}{1 - \dfrac{T^i}{T^o}} \right]_{\Delta G = 0, \, \Delta \mu_{Ca/H} = 0} = - \left[\frac{J'^i_q}{1 - \dfrac{T^i}{T^o}} \right]_{\Delta G = 0, \Delta \mu_{Ca/H} = 0} \tag{34}$$

The coefficient is the thermal conductivity of the protein itself, since we have assumed that the transport is one-dimensional. We have also used the condition that r is negligible at equilibrium. According to the flux equation (19c), the coefficient can be determined when the reaction is at equilibrium, and when the osmotic force is zero. A temperature difference must be created between the vesicle inside and outside. To accomplish this, one could suspend vesicles equilibrated at a higher temperature in a cold bath and measure how the bath changes its temperature with time. Heat leakage through the lipid part of the membrane must be corrected for. All main coefficients are positive, cf. Table 1.

Table 1 Estimated signs of transport coefficients

j	D_{jr}	D_{jr}	D_{jr}
r	+	+	−
d	+	+	−
q	−	−	+

5.2 The Coupling Between Reaction and Diffusion

The coupling coefficient D_{dr} is the central coefficient in active transport. The coefficient is best defined from

$$\frac{D_{rd}}{D_{dd}} = \left(\frac{J_{Ca}}{r}\right)_{\Delta G=0, \Delta T=0} \tag{35}$$

The value of D_{dr} must be positive, see Table 1. Without a positive value for D_{dr} chemical energy is not converted to osmotic work. The related coefficient, D_{rd} ($D_{dr} = AD_{rd}$), can be independently determined from

$$\frac{D_{dr}}{D_{rr}} = \left(\frac{J_{Ca}}{r}\right)_{\Delta G=0, \Delta\mu_{Ca/H}=0} \tag{36}$$

A reduction in D_{dr} during slip leads to a proportional reduction in D_{rd}. From known values of D_{rr}, D_{dd}, r and J_{Ca}, we can also calculate D_{rd} and D_{dr} for the slipping pump.

5.3 The Coupling of Reaction and Diffusion to Heat Flow

There are four coefficients that describe the coupling between reaction or diffusion and heat flow. Only two of them are independent, thanks to the Onsager relations.

The coupling coefficient for the heat flux and the reaction rate can be determined from such measurements:

$$D_{qr} = -\left[\frac{J_q'^i + r\Delta H}{1 - \exp(-\Delta G/RT)}\right]_{\Delta\mu_{Ca/H}=0,\Delta T=0} \tag{37}$$

The equation explains that the rate r and the corresponding value of $J_q'^i$ must be measured for a given value of ΔG and ΔH in an uncoupled, isothermal pump, with $\Delta\mu_{Ca/H} = 0$. As mentioned already, De Meis and coworkers [16] measured the heat flow into the solution during ATP hydrolysis using isothermal calorimetry for various conditions. The measurable heat flux into the solution is given by (18). The coefficient D_{qr} can be calculated from this equation, with knowledge of the variables involved.

The value of the reciprocal coefficient, on the other hand, is found from

$$D_{rq} = -\left[\frac{r}{1 - \dfrac{T^i}{T^o}}\right]_{\Delta\mu_{Ca/H}=0,\Delta G=0} \tag{38}$$

In order to test the measurements for consistency, one can apply the relation (21), when A is known.

The coupling coefficient for the total heat and diffusional flux is defined at chemical equilibrium in an isothermal system. In that situation $r = 0$, and $\Delta H = T\Delta S$. We obtain

$$D_{qd} = -\left[\frac{J_q'^i}{\dfrac{\mu_{Ca/H}^o}{T^o} - \dfrac{\mu_{Ca/H}^i}{T^i}}\right]_{\Delta G=0,\Delta T=0} \tag{39}$$

The value of the reciprocal coefficient is now

$$D_{dq} = -\left[\frac{J_{Ca}}{1 - \dfrac{T^i}{T^o}}\right]_{\Delta\mu_{Ca/H}=0,\Delta G=0} \tag{40}$$

It is reasonable to name all of these coefficients by the name thermogenesis coefficients. They describe the impact of the chemical reaction on the heat flow (D_{qr}) or the effect of a temperature jump on the metabolic rate (D_{rq}). The second set describe the impact of the Ca-flux on the heat flow (D_{qd}), or the effect of a temperature jump on the Ca-flux (D_{dq}).

Experiments indicate that these coupling coefficients are all negative: A negative reaction rate was promoted with $\Delta\mu_{Ca} = 0$ by quenching a warm vesicle suspension to a low temperature [14]. This observation makes D_{qr} and D_{rq} negative. The fact that certain modes of operation are endothermic [17] can be explained if D_{rd} and D_{dr} are also negative. The two first terms of the flux equation for the total heat flux will then have opposite signs, and may shift being the leading term, depending on the conditions.

The signs of all coefficients are summarised in Table 1.

6 Conclusion

We have given a kinetic description of active transport on thermodynamic grounds. We have derived the description from the entropy production of a molecular pump, the Ca-ATPase, in particular the entropy production for events that take place as the reaction proceeds along the reaction coordinate. Flux equations were obtained in the systematic manner characteristic for non-equilibrium thermodynamics. When the description is integrated to the level of observation, non-linear flux force relations were found – as new flux was taken along the total heat flux into the solution. Doing this, we are able to define thermogenesis in terms of four thermogenesis coefficients, of which two are dependent. As the pump starts to slip, the heat production increases. An experimental determination of the transport coefficients was discussed.

Acknowledgements The authors are grateful to the Storforsk grant from the Norwegian Research Council.

References

1. P. Mitchell: Nature, **191**, 144 (1961)
2. L. Onsager: Phys. Rev., **37**, 405 (1931)
3. L. Onsager: Phys. Rev., **38**, 2265 (1931)
4. A. Katchalsky and P. Curran: *Nonequilibrium Thermodynamics in Biophysics* (Harvard University Press, Cambridge, Massachusetts, 1975)
5. S.R. Caplan and A. Essig: *Bioenergetics and Linear Nonequilibrium Thermodynamics. The Steady State* (Harvard University Press, Cambridge, Massachusetts, 1983)
6. H.V. Westerhoff and K. van Dam: *Thermodynamics and Control of Biological Free-energy Transduction* (Elsevier, Amsterdam, 1987)
7. D. Walz: Biochim. Biophys. Acta, **1019**, 171 (1990)
8. D. Bedeaux: Adv. Chem. Phys. **64**, 47 (1986)
9. S. Kjelstrup, D. Bedeaux and E. Johannessen: *Elements of Irreversible Thermodynamics for Engineers*, 2nd. ed. (Tapir Akademiske Forlag, Trondheim, Norway, 2006)
10. D. Reguera, J.M. Rubi and J.M.G. Vilar: J. Phys. Chem. B, **109**, 21502 (2005)
11. S. Kjelstrup, J.M. Rubi and D. Bedeaux: J. Theor. Biology, **234**, 7 (2005)
12. S. Kjelstrup, J.M. Rubi and D. Bedeaux: Phys. Chem. Chem. Phys. **7**, 4009, (2005)
13. L. de Meis and R.K. Tume: Biochemistry **16**, 4455 (1977)
14. L. de Meis and G. Inesi: J. Biol. Chem., **257**, 1289 (1982)

15. L. de Meis, M.L. Bianconi, V.A. Suzano: FEBS Letters, **406**, 201 (1997)
16. L. De Meis: J. Biol. Chem.: **27**, 25078 (2001)
17. L. de Meis: J. Membr. Biol. **188**, 1 (2002)
18. L. de Meis, A.P. Arruda, R. Madeiro de Costa and M. Benchimol: J. Biol. Chem., **281**, 16384 (2006)
19. H. Eyring and E. Eyring: *Modern Chemical Kinetics* (Chapman and Hall, London, 1965)
20. C. Peinelt and H.J. Apell: Biophysical J., **86**, 815 (2004)
21. C. Toyoshima and G. Inesi: Ann. Rev. Biochem. **73**, 269 (2004)
22. S.R. de Groot and P. Mazur: *Non-equilibrium Thermodynamics* (Dover, New York, 1984)
23. A.R. Waldeck, K. van Dam, J. Berden et al.: Eur. Biophys. J., **27**, 255 (1998)
24. J. Ross and P. Mazur: J. Chem. Phys. **35**, 19 (1961)
25. H.A. Kramers: Physica, **7**, 284 (1940)
26. A.M. Albano and D. Bedeaux: Physica A, 1987, **147**, 407 (1987)
27. A. Røsjorde, D.W. Fossmo, S. Kjelstrup et al.: J. Colloid Interf. Sci., **232**, 178 (2000)
28. E. Johannessen and D. Bedeaux: Physica A, 2003, **330**, 354 (2003)
29. T.L. Hill: *Thermodynamics of Small Systems* (Dover, New York, 1994)
30. T.L. Hill: *Free Energy Transduction and Biochemical Cycle Kinetics*, (Springer Verlag, New York, 1989)
31. K.S. Førland, T. Førland, S. Kjelstrup: *Irreversible Thermodynamics. Theory and Applications*, 3rd. ed. (Tapir Akademiske Forlag, Trondheim, Norway, 2001)
32. L. Torner and J.M. Rubi: Phys. Rev., A **44**, 1077 (1991)
33. I. Antes, D. Chandler, H. Wang et al.: Biophysical J., **85**, 695 (2003)
34. W.S. da-Silva, F.B. Bomfim, A. Galina et al.: J. Biol. Chem., **279**, 45613 (2004)
35. M.C. Berman: Biochim. Biophys. Acta, **1513**, 95 (2001)
36. E.M. Diamond, B. Norton, D.B. MacIntosh et al.: J. Biol. Chem., **255**, 11351 (1980)
37. S.X., Sun, H. Wang, and G. Oster: Biophysical J., **86**, 1373 (2004)

Energy Interconversion in Transport ATPases

Role of Water in Ions Transport and in the Energy of Hydrolysis of Phosphate Compounds

L. de Meis

Abstract This chapter is related to the work carried out by Kjelstrup et al. [1, 2, 3], describing the energy dissipation of uncoupled and coupled enzymes. Here we describe the biochemical experiments that led to a partial understanding of how energy is handled by enzymes (proteins) to transport ions across a biological membrane and how the enzyme is able to determine how much of the total energy available during transport is used to perform work (ion transport) and how much is dissipated as heat. The experiments described show that the water organized around proteins (enzymes) and reactants involved in the transport process play a key role in the mechanism of energy transduction. Most of the bibliography of this chapter is related to the biological experiments that contributed to the elucidation of the mechanism of energy transduction during Ca^{2+} transport, as, e.g., works pertinent to the thermodynamic process of active transport, as seen in the perspective of physics [1, 2, 3, 4, 5, 6, 7].

1 Introduction

In order to survive, living organism developed molecular mechanism that allows the interconversion of different forms of energy. The amount of energy continuously interconverted by living matter is amazingly huge. On a gram per gram basis, a living human tissue converts 10,000 more energy per second than the Sun. Some bacteria, such as *Azotobacter*, can convert 50 million times more energy per second than the Sun [4].

There are a few key features that help the understanding of the process of energy interconversion in animals. These are as follows:

(1) *Phosphate Compounds of "High" and "Low" Energy.*
   Energy is carried through different regions of the cell by phosphate compounds referred to as "high energy phosphate compounds." Among these molecules,

L. de Meis

Instituto de Bioquímica Médica, Prédio do CCS, Universidade Federal do Rio de Janeiro, Cidade Universitária, Rio de Janeiro, RJ, 21941-590, Brazil, demeis@bioqmed.ufrj.br and leodemeis@terra.com.br

de Meis, L.: *Energy Interconversion in Transport ATPases: Role of Water in Ions Transport and in the Energy of Hydrolysis of Phosphate Compounds.* Lect. Notes Phys. **752**, 175–187 (2008)
DOI 10.1007/978-3-540-78765-5_8   © Springer-Verlag Berlin Heidelberg 2008

adenosine triphosphate (ATP) is the most important of them and is thought to be the biological currency of energy exchange. Energy is made available to the system when the phosphate of high energy is cleaved. In the case of ATP, the energy becomes available for interconversion when the γ-phosphate of the molecule is hydrolyzed [5].

$$ATP + H_2O \leftrightarrow ADP + P_i + chemical\ energy$$

This equation is found in any biochemistry textbooks. What is not written in textbooks is that, up to present, we do not know how and the way the cleavage of ATP releases energy.

(2) *Enzymes that are Able to Interconvert Energy.*

Cells have a variety of enzyme (ATPases) which catalyze ATP hydrolysis and simultaneously convert the chemical energy made available into work as, for instance, actomyosin during muscle contraction, ion transport mediated by the Ca^{2+}-ATPase and the $Na^+ + K^+$-ATPase, formation of gradients through membranes (osmotic energy), synthesis of new proteins that must be replaced, etc. Not all the energy derived from the hydrolysis of phosphate compounds is converted by enzymes into work; a part of the total energy released is dissipated in the surrounding environment as heat. We do not know how proteins (enzymes) handle the energy released during a chemical reaction and interconvert it in work or other forms of energy [6, 7].

(3) *Homeothermy*

Humans and a large array of animal are homeothermic, i.e. they are able to maintain the body temperature within a narrow range, usually oscillating between 35°C and 38°C, regardless of the large environmental temperature changes. Homeothermy is thought to be an important evolutionary acquisition because it allows optimal conditions for biological reactions regardless of ambient temperature. There is a sophisticated system in the brain that control the intake of energy (appetite) and the dissipation of energy as heat into the environment [8]. We know that hormones play a role in these brain regions. These include insulin synthesized in the pancreas, thyroid hormone T_3, and leptins produced by adipose tissue. However, we do not know yet how this is orchestrated by the brain and what are the mechanisms used by the body cells to respond to the brain command to produce more or less heat.

(4) *Diseases*

The maintenance of the body temperature (homeothermy) requires a sophisticated balance between the rates of heat production and heat dissipation. Alterations of the heat production rate are observed in hormonal dysfunctions [8, 9, 10, 11, 12, 13, 14]. Alteration of body weight are intimately associated with the rate of heat dissipation by the organism, a decrease of it leads to obesity and a highly increased rate of heat production lead to loss of weight and even to cachexia associated with various diseases including cancer, AIDS, and tuberculosis.

2 Concepts in Energy Transduction

Until recently it was generally assumed, first, that the energy derived from hydrolysis of phosphate compounds is only liberated at the precise moment of the phosphate bond hydrolysis and, second, that the amount of energy released and the fraction of it that is converted into heat is always the same regardless of whether the compound is cleaved in solution or on the enzyme surface [5, 7]. In other words, enzymes would not be able to regulate the conversion of the chemical energy released during hydrolysis into either work or heat. These two concepts are no longer valid. We now know that the energy of phosphate compounds' hydrolysis varies greatly depending on whether it is in solution or bound to the enzyme [15, 16]. Moreover, for enzymes involved on energy transduction, the energy used to perform work becomes available *before* cleavage of the phosphate compound (Table 1). Recently we found that enzymes can modulate the conversion of energy during catalysis determining the fraction of energy released which will be converted into work and the fraction to be converted into heat.

Table 1 Variability of the energy of hydrolysis of phosphate compounds during the catalytic cycle of energy transducing enzymes

Enzyme	Reaction	In solution or enzyme bound, *before work*		Enzyme bound, *after work*	
		K_{eq} (M)	ΔG° Kcal/mol	K_{eq} (M)	ΔG° Kcal/mol
Ca^{2+}-ATPase and N^+/K^+-ATPase	Aspartyl phosphate hydrolysis	10^6	-8.4	1.0	0
F_1-ATPase and Myosin	ATP hydrolysis	10^6	-8.4	1.0	0
Inorganic Pyrophosphatase	PP_i hydrolysis	10^4	-5.6	4.5	-0.9
Hexokinase	ATP + gluc $\rightarrow$ gluc-P + ADP	2×10^3	-4.6	1.0	0

For details, see [12] and [23]. The relationship between the standard free energy of hydrolysis (ΔG°) and the equilibrium constant of the reaction (K_{eq}) is $\Delta G^\circ = RT \ln K_{eq}$, where R is the gas constant (1.981) and T is absolute temperature.

3 Phosphate Compounds of High and Low Energy

The history of ATP dates back to the late 1920s when Fiske and Subbarow [17] were searching for a method that allowed the measurement of inorganic phosphate (Pi) in animal tissues. During the course of their experiment, they purified ATP

and creatine phosphate from animal tissues. At that time the physiological importance of these phosphate compounds was not known. It took several years to discover that ATP was an energy carrier molecule and that its hydrolysis provides
the energy needed for muscle contraction. It was only in 1941 that the concept of
"energy-rich" and "energy-poor" phosphate compounds was formally presented by
Lipmann [5] in a review where he analyzed the data obtained in his and other laboratories and on the basis of the knowledge available at that time Lipmann proposed the
following:

(1) Energy-rich phosphate compound would be those who presented a high equilibrium constant (K_{eq}) for the hydrolysis of the phosphate attached to the molecule
and, conversely, low energy phosphate compounds would have a low equilibrium
constant of hydrolysis. In this view, phosphoesters such as glucose-6-phosphate and
glycerol phosphate would be typical phosphate compounds of low energy and ATP
a high-energy molecule.

(2) The energy that could be derived from the hydrolysis of a phosphate compound would depend exclusively on the chemical nature of the bond that links
the phosphate residue to the rest of the molecule. The N–P bond of creatine phosphate, the enol phosphate of phosphoenolpyruvate, the phosphoanhydride linkages
of ATP and pyrophosphate (PPi), and the acyl phosphate bond of aspartyl phosphate
were typical energy-rich phosphate bonds which have K_{eq} values much higher than
low-energy phosphate such as glucose-6-phosphate. These concepts remained unchanged from the time of Lipmann review in 1941 until the 1970s. During this
period, most work was of a theoretical nature and the "high-energy" nature of
the phosphate bond was thought to be dependent on intramolecular effects such
as opposing resonance, electrostatic repulsions, and electron distribution along the
molecule backbone [18, 19, 20]. Thus, the high K_{eq} for the hydrolysis of pyrophosphate and ATP was thought to be determined by the negative charges and opposing
resonance of the P–O–P linkage. The negative charge on either side of the linkage
would repel each other, creating tension within the molecule and the opposing resonance would generate points of weakness along the backbone. As a result, it would
be easy to cleave the molecule and difficult to bring together the products of the
hydrolytic reaction. The combination of these two factors would be responsible for
the high equilibrium constant (K_{eq}) for the hydrolysis ATP or PPi phosphoanhydride bonds. In 1970 George and co-workers [21] reasoned that in biological systems phosphate compounds are in solution and, thus, interact with water. It will
be, therefore, expected that water molecules should organize themselves around
the phosphate compound and this would both shield the charges of the molecule,
thus neutralizing the electrostatic repulsion and, in addition, form bridges between
different atoms of the molecule, thus reinforcing the weak points generated along
the molecule backbone by opposing resonances. George and co-workers [21] proposed that the K_{eq} for the hydrolysis of a phosphate compound should be determined by the differences in solvation energies of reactants and products and not by
intramolecular effects as previously proposed. Solvation energy is the amount of
energy needed to remove the solvent molecules that organize around a substance
in solution. A more solvated molecule would then be more stable or less reactive

than a less-solvated molecule and the k_{eq} for hydrolysis of a phosphate compound would be determined by the difference of solvation energy between reactants and products. When lower the solvation energy of the products, higher would be the K_{eq} of the reaction. According to George et al.'s calculations, the solvation energy of Pi varies between 76 and 637 kcal /mol and for PPi between 87 and 584 kcal/mol. The variability of the values is related to the pH of the medium that determines the different degrees of ionization of the two molecules. When more alkaline the medium, more ionized and more solvated are Pi and PPi. The measured free energy of hydrolysis of PPi in water solution varies between -4 and -6 Kcal/mol. This represents a very small fraction of the total solvation energy of either Pi or PPi. Thus, a small change in the organization of the solvent around the molecules of reactants and products would be sufficient to account for the measured energy of hydrolysis of pyrophosphate. The work of George and co-workers remained unnoticed for several years. It was only in 1978 that the solvation theory was revised and substantiated by Hayes and coworkers [22]. These authors calculated the energy of hydrolysis of several phosphate compounds in gas phase, a theoretical situation in which reactants and products are no longer solvated and compared the values estimated with those measured in aqueous solutions. These calculations implied significant approximations. Experimentally, gas phase is only reached at very high temperatures, various orders of magnitude higher that the temperature in which measurements in aqueous solutions was made. Furthermore, at temperatures similar to those needed to reach gas phase ATP, phosphocreatine and other organic molecules are decomposed. According to Hayes et al.'s calculations [22], in aqueous solution, acetyl phosphate and the N–P bonds of phosphocreatine and phosphoarginine are of a high-energy nature. However, in gas phase this would no longer be true. On the contrary, positive free energy of hydrolysis calculated indicated that when reactants and products are not solvated, acetyl phosphate and phosphocreatine are more stable than the products of their hydrolysis, and according to Lipmann's definition [5], they behave as "energy poor" phosphate compounds. Interestingly, the concept of solvation energy seems to be valid only for the "high-energy" and not for the "low-energy" phosphate compounds because the values calculated for the energies of hydrolysis of glucose-6-phosphate and others phosphoesters were the same regardless of whether the molecules were in aqueous solutions or in gas phase [21, 22].

4 Experimental Measurements

Starting in 1984 we became interested in the solvation theory. The motivation was the discovery that the same chemical specie, an acyl phosphate residue, would have different energies of hydrolysis depending on whether it was in solution (water) or on the enzyme surface. This will be discussed below in a different section. In order to test the solvation theory experimentally, we measured in our laboratory the energy of hydrolysis of PPi in media with different water activities. The strategy adopted was to measure the K_{eq} in aqueous media containing different concentrations of organic solvents which are known to decrease the water activity (w_a) without changing the dielectric constant of the solution. These measurements were performed at physiological temperature (35°C) and pH values (pH 6.0 up to pH 8.0). The K_{eq} values we measured were different from those theoretically calculated, but in agreement with the solvation theory, we found that a discrete change of water activity could promote a significant change of PPi energy of hydrolysis (Fig. 1). In totally aqueous media ($w_a = 1$) the $\Delta G°$ of PPi hydrolysis was −5 kcal/mol, but it raised to +1 kcal/mol when the w_a was decreased to 0.5 [15, 23, 24, 25, 26]). Thus, depending on the w_a of the medium, PPi could change from a "high energy" to a "low energy" compound. One year later Williams and Wolfenden [27] found that ATP could be spontaneously synthesized in wet chloroform and recently Kirby et al. [28] observed that acetyl phosphate can be spontaneously synthesized in a media with a decreased water activity. In agreement with Hayes and coworkers calculations [22] we found that w_a has practically no effect on the energy of hydrolysis of phosphoesters such as glucose-6-phosphate (Fig. 1). Thus, in totally aqueous medium, PPi behaves as a high-energy compound when compared to glucose-6-phosphate, but in an hydrophobic environment ($w_a = 0.5$) glucose-6-phosphate becomes a phosphate compound with a higher energy of hydrolysis than pyrophosphate [8, 22]. These effects of w_a on the free energy of hydrolysis of PPi and glucose-6-phospahte can be easily interpreted according to the solvation energy theory [21, 22] but difficult to accommodate with the 1941 Lipmann proposal [5]. If the energy of the reaction should be determined solely by the nature of the phosphate bond cleaved as proposed in the early 1940s, then the energy of hydrolysis of PPi should not depend on the w_a of the medium and always be higher than that of glucose-6-phosphate regardless of the conditions used. It is known that the w_a along the surface of a protein may vary substantially and it is generally different from the bulk solution in which the protein is solubilized. Solvation energy was found to play an important physiological role in biological processes of energy transduction. One example is the photosynthetic synthesis of PPi from 2 Pi molecules by the chomatophores of the photosynthetic bacteria R. rubrum. The chomatophores have a membrane-bound

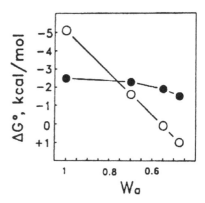

Fig. 1 Effect of water activity in the energies of hydrolysis of PPi (*open symbols*) and glucose-6-phosphate (*closed symbols*). For experimental details see [20, 21, 22]

pyrophosphatase which catalyze the synthesis of PPi when exposed to light and hydrolyze the PPi previously synthesized when the light is turned off (dark). This cycle synthesis and hydrolysis can be mimicked by a soluble pyrophosphatase, without the need of either membrane, chlorophyll, or light but by simply decreasing the w_a of the solution with organic solvent and then diluting the solvent by the addition of water [29, 30].

5 Energy Transduction by Enzymes: Conversion of Phosphate Bonds from High into Low Energy at the Catalytic Site of Enzymes

In the early 1970s, little was known about mechanisms of energy transduction by enzymes and it was thought that the energy of hydrolysis of phosphate compounds would be the same regardless of whether they were bound to the enzyme or free in solution. For enzymes which transduce energy as, for instance, an ATPase capable of transporting ions across a membrane, it was thought that the region of the protein where the catalytic site was located should be in close vicinity with the region of the protein which would translocase the ion. This would facilitate the transfer of energy between the two parts of the protein without a significant loss of energy and the sequence of events taking place during the process of energy transduction was thought to be as follows:

(1) The enzyme binds ATP;
(2) ATP is hydrolyzed and energy is released at the catalytic site at the precise moment of the phosphate bond cleavage;
(3) The energy is immediately absorbed by the enzyme and used to translocated the ion through the membrane and
(4) The products ADP and Pi dissociate from the enzyme.

For the synthesis of ATP from ADP and Pi, the sequence of events would be the same, but in the reverse order [15, 31, 32].

For several years we have been working in laboratory on the mechanism by which enzymes use the energy derived from ATP or PPi to transport ions across a biological membrane; in particular, on Ca^{2+} transport. The sarco/endoplasmic reticulum Ca^{2+}-ATPases (SERCA) is a family of isoenzymes that can interconvert different forms of energy. Hasselbach and Makinose in 1961 [32, 33, 34, 35] were the first to describe the Ca^{2+} transport in vesicles derived from the sarcoplasmic reticulum of skeletal muscle. These vesicles retain embedded in the membrane a SERCA 1 that uses the energy derived from ATP hydrolysis to accumulate Ca^{2+} in the vesicles lumen. In this process the enzyme converts the chemical energy (ATP hydrolysis) into osmotic energy given by the Ca^{2+} gradient formed across the vesicles membrane. Later, Makinose [33] demonstrated that catalysis is initiated by phosphorylation of and aspartyl residue located in the catalytic site of the enzyme forming an acyl phosphate residue that in water has the same energy of hydrolysis as ATP and, in fact, this reaction is fully reversible (K_{eq} ~1). Energy for the transport of Ca^{2+} across the membrane would only become available during the hydrolysis of the acyl phosphate residue intermediate. The Ca^{2+}transport would then be defined by the following sequence of reactions:

$$E + ATP \leftrightarrow E \sim P + ADP; \textbf{(2)} \; E \sim P + H_2O \rightarrow E + Pi + \textit{energy} \qquad (1)$$

where E is the enzyme and E~P is the enzyme phosphorylated by ATP. In the 1960s Peter Mitchell [36] proposed the chemiosmotic theory to explain the mechanism of ATP synthesis by mitochondria. According to Mitchell the energy derived from oxygen consumption and electron flow through the mitochondrial cytochrome would be used to form a proton electrochemical gradient between the inner membrane and the mitochondrial matrix. The energy derived for the H^+ gradient would then be used by the F_1-F_0 ATPase embedded in the mitochondrial inner membrane to synthesis of ATP from ADP and Pi. In absence of respiratory substrate, the very same enzyme can hydrolyze ATP and use the energy released to form a H^+ gradient. Thus the F_1-F_0 ATPase, now called ATP synthase, is able to interconvert osmotic and electrical energy derived from the H^+ gradient into chemical energy. Based on the Mitchell theory, Hasselbach and Makinose [37, 38, 39] demonstrated that the SERCA was also able to use the energy derived from an ionic gradient to synthesize ATP. During synthesis, the entire process of Ca^{2+} transport is reversed and the two reactions above should flow in the reverse direction:

$$E + Pi + \textit{energy} \; E \leftrightarrow E \sim P + H_2O; \textbf{(1)} \; E \sim P + ADP \rightarrow ATP + E \qquad (2)$$

This was shown filling vesicles derived from muscle reticulum with Ca^{2+} and then diluting them in a media containing a Ca^{2+} chelating substance (EGTA), thus forming a steep Ca^{2+} gradient across the vesicles membrane. In these conditions, the Ca^{2+} retained by the vesicles was released to the medium at a slow rate. However, when the end products of ATP hydrolysis, i.e., ADP and Pi, were included in the medium, the Ca^{2+} contained in the vesicles was rapidly released and coupled with the Ca^{2+} efflux; ATP was synthesized from ADP and Pi. Shortly after, Makinose [40] demonstrated that during reversal, the very same aspartyl residue phosphorylated by ATP during Ca^{2+} accumulation was now phosphorylated by Pi forming the same acyl phosphate residue at the catalytic site of the SERCA. Because this intermediate has a high energy of hydrolysis in water, it was thought that the energy derived from the gradient was absorbed by the enzyme to form the aspartyl phosphate intermediate and then the E~P intermediate would transfer its phosphate spontaneously to ADP forming ATP without a further energy need. In 1972, we observed that the SERCA could be phosphorylated by Pi *in the absence* of a Ca^{2+} gradient, i.e., in the absence of an apparent source of energy [15, 32, 41, 42]. This was achieved incubating leaky vesicles in a medium without Ca^{2+}, i.e., the same medium that the outer surface of the vesicles is exposed when a gradient is formed across the vesicles membrane. This phosphoenzyme, however, was of "low energy" because it could be formed spontaneously ($K_{eq} = 1$; $\Delta G^\circ = 0$) and was not able to transfer its phosphate to ADP to form ATP. The "low energy" phosphoenzyme was converted into "high energy" when the leaky vesicles were transferred from a medium without Ca^{2+} to a medium containing a high Ca^{2+} concentration, similar to that found in the vesicles lumen when a gradient was formed [43, 44, 45]. The phosphoenzyme was now able to transfer its phosphate to ADP forming ATP in the absence of a Ca^{2+} gradient. Thus, we could complete a single catalytic cycle of ATP synthesis exposing the enzyme first to zero Ca^{2+} and then to a medium with high Ca^{2+} concentration. Subsequent kinetics experiments demonstrated that the transition from "low" to "high energy" was associated with a conformational change of the protein during which Ca^{2+} was translocated through the membrane [15, 16, 32, 46]. Therefore, the same chemical specie had different energies of hydrolysis depending on the conformational state of the protein and the sequence of intermediary steps during catalysis would then be

$$
\begin{array}{ccc}
& (1) & \\
& E_1 + ATP \leftrightarrow E_1 \sim P + ADP & \\
(4) \updownarrow & \updownarrow \rightarrow energy & (2) \\
& \underline{E_2 + Pi} \leftrightarrow E_2 - P & \\
& (3) &
\end{array}
$$

where E_1 and E_2 are two distinct conformational forms of the enzyme and $E_1 \sim P$ and E-P are phosphoenzyme of "high" and "low" energies respectively. During Ca^{2+} uptake the sequence would flow from reaction 1 to 4 and during reversal of the catalytic cycle the direction of the intermediary reactions would be inversed and flow from reaction 4 to 1 leading to ATP synthesis. Based on the concepts of solvation energy described above, we raised the possibility that the transition from E~P of high energy into E-P of low energy would be associated with a change of water activity at the catalytic site of SERCA. This possibility was substantiated with the use of non-denaturizing organic solvent such as dimethyl sulfoxide. Transition from high into low energy could be achieved by simply changing the water activity of the medium, and in the absence of a gradient the enzyme was able to complete a catalytic cycle synthesizing ATP solely by changing the water activity of the medium. The findings described above by us were confirmed in different laboratories using different experimental approaches [15, 47, 48]. Toyoshima [49, 50] was able to crystallize the two conformations of SERCA 1. In the E_1 conformation the catalytic site is open and exposed to the medium while in the E_2 conformation the catalytic site is occluded and the interior of the site has very little access to the surrounding solvent. The finding obtained with SERCA and the role of water activity in the conversion of phosphate compound from high into low energy was extended to other enzymes. The same year we observed the formation of the low-energy SERCA phosphoenzyme; in Boyer laboratory [51] it was found that the mitochondrial F_1 ATPase was able to catalyze a rapid phosphate–oxygen exchange in the absence of H^+ gradient. These experiments indicated that ATP could be synthesized at the catalytic site of the mitochondrial ATPase without the need of the energy derived from the proton gradient. Later, kinetic evidences indicated that the F_1 ATPase undergoes three different conformational states. In one of them ATP is synthesized spontaneously ($K_{eq} \sim 1$). Energy from the gradient is then needed to change the conformation of the enzyme and for the conversion of ATP from "low" into "high" energy. The use of organic solvents indicated that as for the SERCA, during the conformational changes there is a change of water activity at the catalytic site of the mitochondrial F_1ATPase which propitiates the conversion of ATP from high into low energy [15, 45, 46, 47, 48, 49, 50, 51, 52]. Table 1 shows the different enzymes where phosphate compounds are converted at the catalytic site from high into low energy before cleavage.

According to these findings, the energy for work does not become available to the enzyme at the moment of the phosphate compound cleavage and the sequence of events now proposed for the process of energy transduction is as follows:

(1) The enzyme binds ATP or other phosphate compounds. For SERCA or $Na^+ + K^+$-ATPase, the catalytic site is phosphorylated by ATP forming and acyl phosphate residue of high energy;

(2) The enzyme performs work without the phosphate compound being hydrolyzed. This is accompanied by a simultaneously decrease of water activity in the catalytic site (hydrophilic-hydrophobic transition) and by a decrease of the phosphate compound energy level; and

(3) The phosphate compound is cleaved and the products of hydrolysis dissociate from the enzyme in a process which does not involve energy change.

For the synthesis of ATP from ADP and Pi, the sequence of events would be the same, but in the reverse order. In this sequence, most of the energy is used to promote a significant conformational change of the enzyme (conformational energy) which in turn promotes a change of water organization along the protein (salvation energy). The transference of energy is carried out by water molecules (salvation energy) which dissociate from a protein region (catalytic site) and binds to another part of the enzyme where work is to be performed. The conformational change is therefore needed to occlude and squeeze out water molecules from a part of the protein and, simultaneously, to expose other parts of the protein allowing the binding of water. After work performance and release of the products of the substrate used, the enzyme returns to the initial conformation in order to start a new catalytic cycle.

6 Energy Transduction and Heat Production in Transport ATPases

In reactions involving energy transduction, only a part of the chemical energy released during the hydrolysis of ATP is converted into work or other forms of energy such as osmotic energy. The other part is converted into heat, and in endothermic animals the heat released is used to maintain a constant and high body temperature.

Interest in heat production and thermogenesis has increased during the past decade due to its implications in health and disease. Alterations of thermogenesis are noted in several disorders, such as control of body weight and endocrine dysfunction. The thyroid hormone T_3 (3,5,3'-tri-iodo-L-thyronine) is involved in the thermal regulation of vertebrates, and in hyperthyroidism there is a decrease of body weight, and an increase of both the basal metabolism and the rate of heat production [53, 54, 55, 56, 57, 58].

Until a few years ago, it was assumed that the amount of heat produced during the hydrolysis of an ATP molecule was always the same and could not be modulated by enzymes, as if the energy released during ATP hydrolysis were to be divided in two non-interchangeable parts: one would be converted into heat and the other used for work. Contrasting with this view, recently it was found that some enzymes are able to handle the energy derived from ATP hydrolysis in such a way as to determine how much is used for work and how much is dissipated as heat [59, 60, 61, 62, 63, 64, 65]. In this view, the total amount of energy released during ATP hydrolysis is always the same, but the fraction of the total energy that is converted into work or heat seems to be modulated by the enzyme. This was first shown for the sarcoplasmic reticulum Ca^{2+}-ATPase [59, 60, 61, 62] and later for hexokinase [65] and plants pyrophosphatase [66].

For the Ca^{2+}-ATPase it was found that the amount of heat released during ATP hydrolysis varies depending on whether the acyl phosphate residue formed by ATP is cleaved before or after its conversion from "high" into "low" energy [59, 60, 61, 62, 63, 67, 68, 69, 70]. Ca^{2+} transport across the membrane takes place during this conversion. If cleavage occurs when the acyl phosphate has a high energy of hydrolysis (hydrophilic environment at the catalytic site) then all the energy derived from the cleavage is converted into heat and none is used for Ca^{2+} transport. Conversely, when the low-energy acyl phosphate is cleaved (hydrophobic environment at the catalytic site) most of the energy was already used for Ca^{2+} transport and little was left available to be converted into heat. Regulation of heat production depends therefore on the ratio between high- and low-energy acyl phosphates residues cleaved during catalysis. There are several SERCA isoforms and the amount of heat produced during ATP hydrolysis and Ca^{2+} transport varies depending on the isoform used. This variability was found to depend on the frequency of the high-energy acyl phosphate residue cleavage. The thyroid hormone T_3 regulates the expression of the various SERCA isoforms. In hyperthyroidism, a condition where the rate of heat production is increased, the SERCA isoforms that produce more heat (SERCA 1) is over expressed. It was shown that the administration of T_4 (a precursor of T_3) to rabbits promotes an increase in the rates of both the uncoupled ATPase activity and heat production in sarcoplasmic reticulum vesicles, in a fashion dependable on the muscle type used [58]. This may suggest that Ca^{2+}–ATPase uncoupled activity may be one of the heat sources contributing to enhanced thermogenesis in hyperthyroidism, and that the theory proposed by Kjelstrup et al. [1, 2, 3], describing the energy dissipation of uncoupled and coupled enzymes, may turn out to be fruitful in a quantitative analysis of these events.

References

1. Kjelstrup, S., Rubi, J. M., and Bedeaux, D. *J. Theor Biol.* **234**, 7–12 (2005).
2. Kjelstrup, S., Rubi, J. M., and Bedeaux, D. *Phys. Chem. Chem. Phys.* **7**, 4009–4018 (2005).

3. Kjelstrup, S. et al.: *A Thermodynamic Description of Active Transport*. Lect. Notes Phys. **752**, 155 (2008).
4. Mitchell, P. *Nature*, **191**, 144 (1961).
5. Reguera, D. Rubi, J. M., and Vilar, J. M. G. *J. Phys. Chem. B*, **109**, 21502 (2005).
6. Hill, T. L. *Thermodynamics of Small Systems* (Dover, New York, 1994).
7. Førland, K. S. Førland, T., and Kjelstrup, S. *Irreversible Thermodynamics. Theory and Applications*, 3rd. ed. (Tapir Akademiske Forlag, Trondheim, Norway, 2001).
8. Schatz, G. *FEBS Lett.*, **536**, 1 (2003).
9. Lipmann, F. *Adv. Enzimol.* **1**, 99–162 (1941).
10. Jansky, L. *Physiol. Rev.* **75**, 237–259 (1995).
11. de Meis, L. *J. Memr. Biol.* **188**, 1–9 (2002).
12. Lowell, B. B. and Spiegelman, B. M. *Nature* **404** (6778), 652–660 (2000).
13. de Meis, L., Arruda A. P. and Carvalho D. P. *Biosci. Rep.* **25**(3–4), 181–190 (2005).
14. Clausen T., Van Hardeveld, C., and Everts, M. E. *Physiol. Rev.* **71**, 733–744 (1991).
15. de Meis, L. *Biochim. Biophys. Acta*, **973**, 333–349 (1989).
16. Wolosker, H., Engelender, S., and de Meis, L. *Adv. Mol. Cell Biol.* **23A**, 1–31 (1998).
17. Fiske, C. H. and Subbarow, Y. *Science* **65**, 401 (1927).
18. Kalckar, M. *J. Biol. Chem.* **137**, 789–790 (1941).
19. Hill, T. L. and Morales, M. H. *J. Am. Chem. Soc.* **73**, 1656–1660 (1951).
20. Boyd, D. B. and Lipscomb, W. N. *J. Theor. Biol.* **25**, 403–420 (1969).
21. George, P., Witonsky, R. J., Trachtman, M., Wu, C., Dorwatr, W., Richman, L., Richman, W., Shuray, F., and Lentz, B. *Biochim. Biophys. Acta* **223**, 1–15 (1970).
22. Hayes M. D., Kenyon, L. G., and Kollman, A. P. *J. Am. Chem. Soc.* **100**, 4331–4340 (1978).
23. de Meis, L. *J. Biol. Chem.* **259**, 6090–6097 (1984).
24. de Meis, L., Behrens, M. I., Petretski, J. H., and Politi, M. *J. Biochem.* **24**(26), 7783–7789 (1985).
25. Romero, P. J. and de Meis, L. *J. Biol. Chem.*, **264**, 7869–7873 (1989).
26. de Meis, L. *Arch. Biochem. Biophys.* **306**, 287–296 (1993).
27. Williams, R. and Wolfenden, R. *J. Amer. Chem. Soc.* **107**, 4345 (1985).
28. Kirby, A. J., Lima, M. F. da Silva, D., and Nome, F. *J. Am. Chem. Soc.* **126**(5), 1350–1351 (2004).
29. de Meis, L., Behrens, M. I., Celis, H., Romero, I., Gomez Puyou, M. T., and Gomez-Puyou, A. *Eur. J. Biochem.* **158**(1), 149–157. (1986).
30. Behrens, M. I. and de Meis L. *Eur. J. Biochem.* **152**(1), 221–227 (1985).
31. Hasselbach, W. *Biochim. Biophys. Acta* **515**, 23–53 (1978).
32. de Meis, L. and Vianna, A. L. *Annu. Rev. Biochem.* **48**, 275–292 (1979).
33. Hasselbach, W. and Makinose, M. *Biochem. Z.* **333**, 518–528 (1961).
34. Hasselbach, W. *Prog. Biophy. Biophys. Chem.* **14**, 167–222 (1964).
35. Inesi, G. *Annu. Rev. Physiol.* **47**, 573–601 (1985).
36. Mitchell, P. and Moyle, J. *Eur. J. Biochem.* **7**, 471–484 (1969).
37. Makinose, M. *FEBS Lett.* **12**, 269–270 (1971).
38. Makinose, M. and Hasselbach, W. *FEBS Lett.* **12**, 271–272 (1971).
39. Hasselbach, W. *Biochim. Biophys. Acta* **515**, 23–53 (1978).
40. Makinose, M. *FEBS Lett*, **25**, 113–115 (1972).
41. Masuda, H. and de Meis, L. *Biochemistry* **12**, 4581–4585 (1973).
42. de Meis, L. and Masuda, H. *Biochemistry* **13**, 2057–2061 (1974).
43. de Meis, L. and Carvalho, M. G. *Biochemistry* **13**, 5032–5038 (1974).
44. Knowles, A. F. and Racker, E. *J. Biol. Chem.* **250**, 1949–1951 (1975).
45. de Meis, L. and Tume, R. K. *Biochemistry* **16**, 4455–4463 (1977).
46. de Meis, L. *The Sarcoplasmic Reticulum: Transport and Energy Transduction*. Ed. E. Bittar. (Wiley, New York, 1981).
47. de Meis, L., Martins, O. B., and Alves, E. W. *Biochemistry*, **19**, 4252–4261 (1980).
48. Tanford, C. *CRC Crit. Revs. Biochem.* **17**, 123–151 (1984).
49. Toyoshima, C. and Mizutani, T. *Nature* **430**, 529–535 (2004)
50. Toyoshima, C., Nomura, H., and Tsuda, T. *Nature* **432**, 361–368 (2004)

51. Sakamoto, J. and Tonomura, Y. *J. Biochem.* (Tokyo). **93**(6), 1601–1614 (1983).
52. Gómez-Puyou, A., Gómez-Puyou, M. T., and de Meis, L. *Eur. J. Biochem.* **159**(1), 133–140 (1986).
53. Boyer, P. D. *Nobel lecture – Biosci. Rep.* **18**, 97–117 (1998).
54. Astrup, A., Buemann, B., Toubro, S., Ranneries, C., and Raben, A. *Am. J. Clin. Nutr.* **63**, 879–883 (1996).
55. Al-Adsani, H., Hoffer, L. J., and Silva, J. E. *J. Clin. Endocrinol. Metab.* **82**, 1118–1125 (1997).
56. Silva J. E. *J. Clin. Invest.* **108**, 35–37 (2001).
57. Andrienko, T., Kuznetsov, A. V., Kaambre, T., Usson, Y., Orosco, A., Appaix, F., Tiivel, T., Sikk, P., Vendelin, M., Margreiter, R., and Saks, V. A. *J. Exp. Biol.* **206**, 2059–2072 (2003).
58. Arruda, A. P., Da-Silva, W. S., Carvalho, D. P. and de Meis, L. *Biochem. J.* **375**, 753–760 (2003).
59. de Meis, L. *Am. J. Physiol.* **274** (*Cell Physiol*).*43*, C1738–C1744 (1998).
60. de Meis, L. *J. Biol. Chem.* **276**(27), 25078–25087 (2001).
61. de Meis, L. *Biosci. Rep.* **21**, 113–137 (2001).
62. de Meis, L. *J. Biol. Chem*, **278**, 41856–41861 (2003).
63. de Meis L., Arruda A. P., da Costa R. M., and Benchimol M. *J. Biol. Chem.* **281**, 16384–16390 (2006).
64. Mall, S., Broadbridge, R., Harrison, S. L., Gore, M. G., Lee, A. G., East, J. M. *J. Biol. Chem.* **281**, 36597–36602 (2006).
65. Bianconi, M. L. *J. Biol. Chem.* **278**, 18709–18713 (2003).
66. da-Silva, W., Bomfim, F. M., Galina, A., and de Meis, L. *J. Biol. Chem.* **279**, 45613–45617 (2004).
67. Yu, X. and Inesi, G. *J. Biol. Chem.* **270**, 4361–4367 (1995).
68. Reis, M., Farage, M., Souza, A. C., and de Meis, L. *J. Biol. Chem.* **276**, 42793–42800 (2001).
69. Barata, H. and de Meis, L. *J. Biol. Chem.* **277**, 16868–16872 (2002).
70. Reis, M., Farage, M., and de Meis, L. *Mol. Membr. Biol.* **19**, 301–310 (2002).

A Novel Mechanism for Activator-Controlled Initiation of DNA Replication that Resolves the Auto-regulation Sequestration Paradox

K. Nilsson and M. Ehrenberg

Abstract For bacterial genes to be inherited to the next bacterial generation, the gene containing DNA sequences must be duplicated before cell division so that each daughter cell contains a complete set of genes. The duplication process is called DNA replication and it starts at one defined site on the DNA molecule called the origin of replication (*oriC*) [1]. In addition to chromosomal DNA, bacteria often also contain plasmid DNA. Plasmids are extra-chromosomal DNA molecules carrying genes that increase the fitness of their host in certain environments, with genes encoding antibiotic resistance as a notorious example [2]. The chromosome is found at a low per cell copy number and initiation of replication takes place synchronously once every cell generation [3, 4], while many plasmids exist at a high copy number and replication initiates asynchronously, throughout the cell generation [5]. In this chapter we present a novel mechanism for the control of initiation of replication, where one type of molecule may activate a round of replication by binding to the origin of replication and also regulate its own synthesis accurately. This mechanism of regulating the initiation of replication also offers a novel solution to the so-called auto-regulation sequestration paradox, i.e. how a molecule sequestered by binding to DNA may at the same time accurately regulate its own synthesis [6]. The novel regulatory mechanism is inspired by the molecular set-up of the replication control of the chromosome in the bacterium *Escherichia coli* and is here transferred into a plasmid model. This allows us to illustrate principles of replication control in a simple way and to put the novel mechanism into the context of a previous analysis of plasmids regulated by inhibitor-dilution copy number control [7]. We analyze factors important for a sensitive response of the replication initiation rate to changes in plasmid concentration in an asynchronous model and discover a novel mechanism for creating a high sensitivity. We further relate sensitivity to initiation synchrony

K. Nilsson
Department of Cell and Molecular Biology, Molecular Biology Programme, Biomedical Centre, SE-75124 Uppsala, Sweden, Karin.Nilsson@icm.uu.se

M. Ehrenberg
Department of Cell and Molecular Biology, Molecular Biology Programme, Biomedical Centre, SE-75124 Uppsala, Sweden, ehrenberg@xray.bmc.uu.se

Nilsson, K., Ehrenberg, M.: *A Novel Mechanism for Activator-Controlled Initiation of DNA Replication that Resolves the Auto-regulation Sequestration Paradox*. Lect. Notes Phys. **752**, 189–213 (2008)
DOI 10.1007/978-3-540-78765-5_9
© Springer-Verlag Berlin Heidelberg 2008

in a synchronous model. Finally, we discuss the relevance of these findings for the control of chromosomal replication in bacteria.

1 Introduction

1.1 Control of Chromosome Replication in E. coli

The regulation of chromosome replication in the bacterium *E. coli* is probably the most extensively studied chromosomal replication control circuit, but it is still far from understood. In *E. coli*, the time of finishing one round of chromosome replication is insensitive to changes in the cell growth rate. It takes approximately 40 min to duplicate the chromosome and the bacterial cell divides approximately 20 min after completion of one round of replication. The shortest generation time of *E. coli* is 20 min and initiation of replication occurs synchronously only once in every cell generation [4, 8]. For cells growing with a generation time smaller than 40 minutes, there are overlapping rounds of replication. Sometimes, synchronous initiation of up to eight *oriC*s must be orchestrated by the control mechanism, to provide one complete chromosome copy to each daughter cell at cell division. Replication starts at the origin *oriC*, from where it proceeds bi-directionally [9]. Initiation takes place when the activator protein DnaA binds to the origin. In addition to the origin, the activator also binds to binding sites scattered along the chromosome [1]. The *datA* locus binds the highest number of activator molecules, amounting to several hundred. The locus is located quite close to *oriC* and is replicated approximately 8 min after initiation of replication [10]. The activator DnaA exists in two forms, DnaA-ATP and DnaA-ADP [1]. During every round of replication there is a regulated inactivation of DnaA-ATP, called RIDA, which converts chromosome-bound DnaA-ATP into DnaA-ADP upon duplication of the DNA molecule [11]. After its synthesis, the DnaA molecule first becomes ATP bound, and only DnaA-ATP can initiate replication [1, 12]. Since, furthermore, only DnaA-ATP can form polymers [13], it is likely that most DnaA molecules bound to the *datA* locus are in the ATP form. Initiation of replication is believed to take place after that the *datA* locus has been saturated by DnaA-ATP molecules. At this point, the concentration of free DnaA-ATP increases sharply, which allows DnaA-ATP molecules to bind to *oriC*, thereby triggering initiation of DNA replication. Following initiation, the *oriC*s become temporarily sequestered, which prevents rapid re-initiation at newly replicated *oriC*s [14]. The length of the sequestration period and the mechanism ending it are not known. Replication converts DNA bound DnaA-ATP into DnaA-ADP and doubles the number of *datA* loci. When the amount of *de novo* synthesized DnaA-ATP is sufficient to saturate the *datA loci* a new round of replication takes place. Both DnaA-ADP and DnaA-ATP can negatively regulate the DnaA synthesis rate by repressing the expression of the *dnaA* gene [15]. Two central questions of chromosome replication control in *E. coli* have remained unanswered, i.e. whether (1) the rate of synthesis of DnaA is regulated and (2) how such a putative regulation can avoid the pitfall of the auto-regulation sequestration paradox [6].

1.2 Control of Plasmid Replication

Plasmids have well-defined copy numbers in their bacterial host cell and encode their own replication control mechanism. Plasmid replication can be categorized into two main groups based on the molecular set-up: anti-sense control, where an inhibitor, a small RNA molecule, negatively controls one or several of the steps in initiation, and iteron control, where iterated DNA sequences located at or near the origin of replication are essential for both initiation and its control [16]. The basic logic of the feedback circuit of anti-sense controlled plasmids (e.g. plasmids ColE1 [17] and R1 [18]) is straightforward. An inhibitor, encoded by the plasmid, negatively regulates the replication initiation frequency per plasmid at rate-limiting step(s) in initiation. The inhibitor is synthesized at a total rate proportional to the plasmid concentration and is unstable, i.e. degraded with a short half-life [7]. In this way, the inhibitor concentration follows the plasmid concentration in the cell. For example, a two-fold increase in plasmid concentration results in a two-fold increase in inhibitor concentration, which reduces the replication frequency per plasmid and vice versa. Initiation of replication is asynchronous in anti-sense controlled plasmids, it takes place throughout the cell generation and the same plasmid origin may fire more than once during a cell generation [5]. In the case of iteron plasmids (e.g. plasmids P_1 [19] and F [20]) the dynamics of the mechanisms of replication control are poorly understood. Whether iteron plasmids initiate asynchronously or synchronously is also not clear [5].

1.3 The Auto-regulation Sequestration Paradox

Binding sites outside the origin of replication and an auto-repressed expression of the gene for an activator of replication, i.e. the activator represses its own synthesis, are two features iteron plasmids share with the chromosomal replication control in E. coli and many other bacteria [1, 20, 21]. This design of the control circuits seems puzzling since the reduction in the free concentration of the activator by sequestration of the binding sites outside the origin is expected to increase the activator synthesis rate by relieving the repression of activator synthesis. Thus, activator sequestration counteracts auto-regulation of the activator synthesis rate, thereby effectively forbidding adequate regulation of initiation of replication. This has been named the *auto-regulation sequestration paradox* [6]. The solution to the paradox in the case of the iteron plasmids P_1 and F seems to be that sequestered activators pair with promoter-bound activators at the start site of activator gene expression by "handcuffing" [22]. Thereby, it is the total, rather than the free, concentration of the activator that auto-regulates expression of the activator gene, and this resolves the *auto-regulation sequestration paradox* [20, 21].

1.4 Sensitivity Amplification Factors

A critical property of any control mechanism is how sensitive its response is to
a variation in its signal. In the case of replication control, the relevant response
to a signal is how sensitively the replication rate per origin responds to changes
in the concentration of origins of replication. The most commonly used measure
of sensitivity is the amplification factor, defined as the percentage change in the
response over the percentage change in the signal [23]. When the response (r) is a
continuous function of the signal (s), the amplification factor is defined as

$$A_{r,s} = \frac{dr}{ds} \cdot \frac{s}{r}. \tag{1}$$

For non-continuous functions, sensitivity amplification over a discrete interval or as
a numerical approximation the amplification factor instead becomes

$$A_{r,s} = \frac{\Delta r}{\Delta s} \cdot \frac{s}{r}. \tag{2}$$

On the intracellular level there are often cascades of signals and responses. The
amplification factor for the final response (r) to a primary signal (s_1) can then be
partitioned into partial sensitivities as

$$A_{r,s_1} = A_{r,s_M} \cdot A_{s_M,s_{M-1}} \cdot \ldots \cdot A_{s_i,s_{i-1}} \cdot \ldots \cdot A_{s_2,s_1}, \quad i = 1,2,\ldots,M. \tag{3}$$

A negative-valued amplification factor means that the signal acts through a negative
feedback mechanism and a positive-valued amplification factor means that the signal acts through a positive feedback mechanism. Often the absolute value of the amplification factor is studied. Then, high sensitivity amplification reflects the ability
to transform a small percentage change in the signal to a large percentage change in
the response. Koshland et al. [24] have classified amplification factors in relation to
hyperbolic control (such as Michaelis–Menten kinetics). They compared the value
of an amplification factor to the value of an amplification factor of the standard
hyperbolic control curve, $r = 1/(1+s/K)$ (negative feedback with response r normalized to zero at infinite signal s) or $r = (s/K)/(1+s/K)$ (positive feedback with
response r normalized to one at infinite signal s), at the same degree of saturation.
For $|A_{r,s}| > 1 - r$ the response to a signal was classified as ultra-sensitive and for
$|A_{r,s}| < 1 - r$ the response to a signal was classified as sub-sensitive.

1.5 Previous Models and Design of Replication Control Circuits

For all replication control circuits it must hold that the concentration of the origin
of replication is regulated by a negative feedback mechanism. An increase in origin
concentration above normal must, in other words, reduce the replication frequency

per origin and vice versa. If the opposite were true and the origin concentration was regulated by a positive feedback mechanism, an increase in origin concentration above normal would lead to a higher and higher replication initiation frequency (runaway replication [7]) and vice versa. In an asynchronous model a negative feedback between origin concentration and replication initiation frequency per origin corresponds to a negative sensitivity amplification factor when calculating the response of the replication frequency per origin to changes in origin concentration at steady state. In a synchronous model all origins initiate replication approximately simultaneously and the origin concentration oscillates approximately two-fold during each round of initiation. Here, steady state is replaced by a steady oscillation of the origin concentration. In a synchronous model a negative feedback mechanism implies that if the origin concentration at a certain point in the oscillation cycle is larger than its concentration at the corresponding point in the steady state cycle, then the inter-replication time will increase and vice versa.

Several models for the replication control of chromosomal replication have been proposed over the years and have been referred to as "negative" and "positive" control mechanisms [25, 26, 27]. In this context the classification into negative and positive control may refer to whether the origin of replication is regulated by an inhibitor or an activator, i.e. whether an inhibitor has to be removed from the origin or if an activator has to bind to the origin, respectively, to activate initiation of replication [27]. However, although some of the models include both possibilities [26], they are still referred to as negative control in the literature. Some models seem to include both negative and positive feedback mechanisms [28], which has created some confusion [29]. As highlighted above, the replication initiation frequency must respond negatively to changes in origin concentration, but the control circuit may contain both positive and negative feedback mechanisms. For instance, an activator may activate initiation of replication by binding to the origin while the synthesis of the activator is negatively regulated by an inhibitor. In an asynchronous model, sensitivity amplification can then be factorized into partial sensitivities, with an odd number of negative-valued sensitivities.

Those early models of chromosome replication control that remain relevant today were formulated in the 1960s. In 1963, Jacob, Brenner and Cuzin proposed positive control in their "replicon" model [25], where an activator binds to the origin of replication and triggers replication of the chromosome (the "replicon"). Pritchard et al. [26] proposed a negative control mechanism, now known as inhibitor-dilution control, in response to the replicon model. They describe how a stable inhibitor with its concentration proportional to the origin concentration can regulate initiation of chromosome replication through a negative feedback mechanism. The model includes (i) a constitutively produced activator triggering replication initiation, (ii) stable inhibitor molecules produced in a fixed number at the time of initiation from a gene located adjacent to the origin or as part of the origin itself and (iii) a two-fold change of the inhibitor concentration to trigger, through a co-operative interaction between the inhibitor and either the origin or the activator, a switch between full and no repression of the origin. In this way, a self-regulated system is created with an inhibitor concentration proportional to the origin concentration that turns off the rate of initiation of replication following initiation by a doubling of the inhibitor concentration. When the volume has doubled by cell growth, so that the inhibitor concentration is halved by dilution, the inhibitor concentration reaches a threshold level that allows replication initiation to be switched on again. Six years after the model by Pritchard et al. [26], inhibitor-dilution control was presented by Fantes et al. [30] in an alternative form with an unstable inhibitor constitutively expressed proportionally to the gene dosage. Later models have attempted to describe the replication control of the *E. coli* chromosome in more detail. In the initiator-titration model, Hansen et al. [28] suggested that (i) binding sites located outside the origin bind the activator with higher affinity than sites at the origin, so that initiation of replication takes place following saturation of all binding sites outside the origin and (ii) that a newly fired origin becomes temporarily refractory to initiation. Both suggestions have turned out to be true [10, 14]. Donachie and Blakely [31] questioned the idea that initiation of replication is triggered above a critical concentration of the activator. Instead they proposed that initiation of replication starts at a certain ratio between active and inactive activators (DnaA-ATP/DnaA-ADP), but they neither specified how the ratio is recognized nor the triggering of initiation of replication. The ratio does vary in a cyclic manner and reaches its highest value at the time of initiation of replication [32], supposedly as a consequence of RIDA, the regulated conversion of DNA-bound DnaA-ATP into DnaA-ADP by replication of the chromosome, and of de novo synthesis of the activator DnaA leading to its DnaA-ATP form.

There is a fundamental difference between gene expression from chromosomes and plasmids, which we suggest has greatly influenced the different designs of chromosomal and plasmid replication control circuits. The concentration of the molecules responsible for gene expression and protein synthesis, the RNA polymerase and the ribosome, respectively, remain approximately constant during the cell cycle. Thus, "during replication of the chromosome, the total rate of gene expression does not change" since it is limited by the total rate of protein synthesis, ultimately limited by ribosome concentration and ribosome efficiency [27]. Thus, the rate of activator synthesis is approximately constant and independent of the chromosome concentration in the cell at a given growth rate. Plasmids, in contrast, are much smaller DNA molecules than the chromosome and one round of replication is therefore completed much faster than one round of chromosome replication. In this case, "*doubling the number of the activator genes on plasmids approximately doubles the total rate of activator gene expression and activator synthesis*" [27].

With a constant activator synthesis rate, independent of the origin concentration of the chromosome a control circuit of replication, initiation may solely consist of a constitutive, unregulated synthesis of an activator that initiates replication by binding to the origin [27]. When the origin concentration rises above normal, the inter-initiation time increases since the concentration of origins to bind and saturate to trigger initiation increases while the rate of activator synthesis stays unchanged and vice versa. However, a constitutive and unregulated synthesis of an unstable inhibitor fails to produce the inhibitor in proportion to the origin concentration. Therefore, a control circuit of replication initiation solely consisting of a constitutively produced inhibitor should not confer replication control [27]. With a synthesis rate proportional to the origin concentration, the opposite is true. So, in the case of plasmids, constitutive expression of an inhibitor, but not of an activator, is expected to generate adequate replication control.

Furthermore, when there is a single copy of the chromosome or an iteron plasmid, there may be large stochastic fluctuations in inhibitor copy numbers. For plasmids with a higher copy number, expression is averaged over more gene copies [33] and constitutive expression of an unstable inhibitor that creates proportionality between the inhibitor and the origin concentration, as suggested by Fantes et al. [30], is commonly used among anti-sense plasmids (see Sect. 1.6).

1.6 Analysis of the Copy Number Control of Plasmid ColE1

Efficient replication control confers small variations in copy number, which minimize plasmid losses at cell division, i.e. the events when a bacterial cell is born without a plasmid copy, and thereby maximize the genetic inheritance of the plasmid. The copy number control mechanism of the anti-sense regulated plasmid ColE1 has been thoroughly analyzed in a deterministic model by Paulsson et al. [34], reviewed in Paulsson and Ehrenberg [7]. The system can be described by the two differential equations

$$\frac{dy}{dt} = k_{Imax} \cdot r_I(s) \cdot y - \mu \cdot y, \tag{4}$$

$$\frac{ds}{dt} = k_s \cdot y - (k_d + \mu) \cdot s, \tag{5}$$

where y is the plasmid concentration and s is the inhibitor (or signal) concentration. The inhibitor regulates the replication initiation rate per plasmid, $k_{Imax} \cdot r_I$. The replication initiation rate per plasmid is the maximal rate constant of replication initiation, k_{Imax} times the degree of activation or the probability that a trial of initiating replication succeeds, r_I, which is a function of the inhibitor concentration, s. The inhibitor molecules are constitutively synthesized by rate constant k_s and actively degraded by rate constant k_d. The plasmid and inhibitor concentration are diluted by cell growth with rate constant μ. The system (4–5) has one non-zero steady state, when $dy/dt = ds/dt = 0$ and the synthesis rate equals the rate of dilution and degradation. Two different scenarios were investigated, i.e. hyperbolic control

$$r_I = \frac{1}{1 + \frac{s}{K_I}} \tag{6}$$

and exponential control

$$r_I = e^{-s/K_I}, \tag{7}$$

where K_I is the apparent dissociation constant of the inhibitor. Hyperbolic control is implemented when the inhibitor controls one rate-limiting step at initiation while exponential control is implemented when the inhibitor controls multiple rate-limiting steps [7, 34]. The quality of replication control depends on the response in the replication initiation frequency per plasmid, $k_{I\max} \cdot r_I$, to changes in the plasmid concentration y. Three parameters have been identified as important for efficient control: (i) the mechanism of inhibition, r_I, (ii) the maximal rate constant of replication initiation, $k_{I\max}$ and (iii) the proportionality between inhibitor (s) and plasmid (y) concentration out of steady state. The mechanism of inhibition determines how the probability of initiating replication changes as a function of the inhibitor concentration. To compare the two inhibition mechanisms, it is instructive to calculate how sensitively the replication initiation frequency changes in response to a change in inhibitor concentration. Since the inhibitor concentration is proportional to the plasmid concentration at steady state in this system (5), the sensitivity amplification factor for the percentage change in $\overline{r_I}$ over the percentage change in plasmid concentration, $\overline{y}$, is the same as the sensitivity amplification factor for the percentage change in $\overline{r_I}$ over the percentage change in inhibitor, $\overline{s}$, at steady state ($\overline{A}_{r_I,y} = \overline{A}_{r_I,s}$), where the bar indicates steady state. The amplification factor $\overline{A}_{r_I,y}$ at steady state is

$$\overline{A}_{r_I,y} = -(1 - \overline{r_I}) \tag{8}$$

for hyperbolic control and

$$\overline{A}_{r_I,y} = \ln \overline{r_I} \tag{9}$$

for exponential control [7]. That is, in the exponential case, the replication initiation rate per plasmid responds ultra-sensitively to the changes in the plasmid concentration for small values of $\overline{r_I}$ and the absolute value of the amplification factor $\left| \overline{A}_{r_I,y} \right|$ monotonically increases with a decreasing value of $\overline{r_I}$ without approaching an upper limit. This behaviour is distinct from the hyperbolic response, where $\left| \overline{A}_{r_I,y} \right|$ approaches the maximal value of 1 when $\overline{r_I}$ approaches zero. Multi-step control is a well-known way to create ultra-sensitive responses [24]. In addition to the mechanism of inhibition, the rate constant of maximal replication initiation, $k_{I\max}$, is an important parameter for efficient replication control. Since $\overline{r_I} = \mu/k_{I\max}$ holds at steady state (4), it is $k_{I\max}$ in relation to μ, the cell growth rate, that determines the degree of activation, $\overline{r_I}$, at steady state [7]. A large value of $k_{I\max}$ corresponds to a small value of $\overline{r_I}$, which gives a high value of $\left| \overline{A}_{r_I,y} \right|$ (within the limit set by the mechanism of inhibition, (8–9)). The amplification factor $\left| \overline{A}_{r_I,y} \right|$ governs the rate by which the plasmid concentration adjusts back to steady state following a deviation in plasmid concentration. For small deviations the adjustment rate constant becomes $k_{\mathrm{adj}} = -\mu \cdot \left| \overline{A}_{r_I,y} \right|$ [34], which is always true if a change in plasmid concentration is described by (4) (see Sect. 2.2). In a stochastic model a high value of k_{adj} corresponds to a low probability that the plasmid copy number is below the steady state number and thereby the probability that the plasmid is lost (that a plasmid-free daughter cell is formed) at cell division is low [35]. The third identified parameter important for efficient replication control is proportionality between the plasmid and the inhibitor concentration outside steady state. This is implicitly assumed in the calculation of the amplification factor $\left| \overline{A}_{r_I,y} \right|$ since it is calculated at steady state, where the inhibitor concentration is always proportional to the plasmid concentration (5). Thus, to transfer a high-absolute value of the amplification factor into a high adjustment rate for correcting deviations in plasmid concentration from steady state, it is required that the inhibitor is degraded by a high rate so that the inhibitor concentration closely follows changes in plasmid concentration. The larger the value of $|\overline{A}_{r_I,y}|$ the larger must be the first-order rate constant of inhibitor degradation [7, 34].

1.7 A Novel Mechanism of Replication Control with Interesting Properties

The molecular set-up of the novel mechanism of replication control contains an activator molecule triggering initiation of replication by binding to the origin of replication. In addition, there is a binding locus outside the origin to which activator

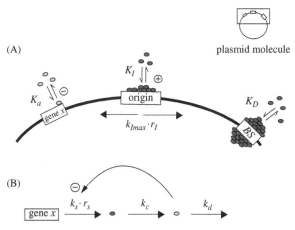

Fig. 1 (**A**) The region around the origin of a hypothetical plasmid regulated by the novel control mechanism. The whole plasmid molecule is shown in the *upper right corner*, with the blown up region indicated. Activator molecules in the active form ("activators") (*filled dark grey ellipses*) bind to the origin with dissociation constant K_I, and trigger the initiation of replication (indicated by a *plus sign*) by rate $k_{Imax} \cdot r_I$. Replication proceeds bi-directionally as shown by the two *arrows* beneath the origin. In addition to the origin, the active activators bind to a binding site (BS) with dissociation constant K_D. Since $K_D \ll K_I$ holds, the activator saturates BS before binding to the origin. Activator molecules in the active form are converted into inactive activators ("inhibitors") (*filled light grey ellipses*). The inactive form of the activator represses gene expression of the activator gene (gene x) by blocking the start site of gene expression (indicated by a *minus sign*) and thereby inhibits the de novo synthesis of activator molecules in the active form. The dissociation constant of inactive activators binding to gene x is denoted as K_a. (**B**) The flow of activator synthesis, conversion and degradation. Activators in the active form are synthesized by rate $k_s \cdot r_s$, where r_s is a function of the concentration of inactive activators. Active activators are converted into activators in the inactive form with rate k_c and inactive activators are degraded by rate k_d

molecules bind with higher affinity than to the origin. The activator is present in two forms, an active and an inactive form. It is only activators in the active form that can bind to the binding site or to the origin and trigger replication (Fig. 1a). The activator is synthesized in the active form and is converted into the inactive form in a first-order reaction. Inactive activators are unstable and are rapidly degraded. The rate of synthesis of activators is negatively regulated by their inactive form (Fig. 1b). The principle for the replication control is straightforward. The large and constant turnover rate per molecule of the activators from their active to their inactive form results in a concentration of the inactive form that is proportional to the *total* concentration of the active form. Thereby, the rate of synthesis of activators can be regulated so that the *total* concentration of the active form remains constant at varying plasmid concentration. Since the activators bind to the sequestration site outside the origin of replication with higher affinity than to the origin, replication is triggered following saturation of the sequestration locus of all plasmids in the cell. Replication events increase the number of sequestration sites, and the cycle of saturating the binding *loci* starts anew. The novel mechanism of replication control, where the activator is converted from an active to an inactive form, offers a new solution to the auto-regulation sequestration paradox. The key is the rapid conversion of the active to the inactive form of the activator molecule, which allows for control of the total, rather than the free, concentration of the activator.

2 Analysis

2.1 Asynchronous Plasmid Model

Inspiration to the new plasmid copy number control mechanism comes from the molecular set-up of the regulation of chromosome replication in the bacterium *E. coli*, where all origins initiate replication synchronously. However, the previous thorough analysis of a replication control mechanism concerns the asynchronously initiating plasmid ColE1 [7, 34]. Since it is not apparent how to relate this analysis to initiation synchrony, we first apply the new regulatory mechanism to plasmids in an asynchronous model to calculate and analyze the central sensitivity parameter $\overline{A}_{r_l,y}$. We thereafter modify the asynchronous plasmid model into a synchronous plasmid model to study the effects of varying the sensitivity parameter also in the synchronous case. This type of analysis has, to our knowledge, not been carried out before. In general, plasmids have much shorter DNA sequences than chromosomes. In the asynchronous case, therefore, we neglect the time to duplicate a plasmid after initiation of replication, meaning that the concentration of complete plasmids is approximated by the concentration of origins of replication. To mechanistically account for initiation synchrony of all plasmids in a cell, we will in that context include a time-delay between initiation and completion of replication.

In the asynchronous model, the plasmid concentration (y) is controlled by the free concentration of the active activator (x_f). The active form of the activator (from now on referred to as the "activator") with total concentration (x_t) is converted to the inactive form with concentration (x_d) (from now on referred to as the "inhibitor"). The inhibitor negatively regulates the synthesis rate of the activator.

$$\frac{dy}{dt} = k_{Imax} \cdot r_I \cdot y - \mu \cdot y = k_{Imax} \cdot \frac{\left(\frac{x_f}{K_I}\right)^{n_I}}{1 + \left(\frac{x_f}{K_I}\right)^{n_I}} \cdot y - \mu \cdot y, \tag{10}$$

$$\frac{dx_t}{dt} = k_s \cdot r_s \cdot y - (k_c + \mu) \cdot x_t = k_s \cdot \frac{1}{1 + \left(\frac{x_d}{K_a}\right)^{n_a}} \cdot y - (k_c + \mu) \cdot x_t, \tag{11}$$

$$\frac{dx_d}{dt} = k_c \cdot x_t - (k_d + \mu) \cdot x_d \tag{12}$$

The replication initiation rate per plasmid in (10) is factorized into the maximal rate, k_{Imax}, and the degree of activation, r_I. The parameter n_I is a co-operativity factor and K_I is the dissociation constant of activator binding to the origin of replication. The activator synthesis rate in (11) is factorized into the maximal rate k_s and the degree of activation of activator gene expression, r_s. The parameter n_a is a co-operativity factor and K_a is the dissociation constant for inhibitor binding. The activator is converted into the inhibitor by the first-order rate constant k_c and the inhibitor is actively degraded by the rate constant k_d. All concentrations in (11–12) are diluted by the cell growth rate μ.

Besides to the origin, the activator also binds to an activator sequestering binding locus (BS) on the plasmid with a dissociation constant K_D, which is much smaller than K_I, meaning that BS must be filled with activator molecules to allow for the increase in free activator concentration that triggers initiation of replication. The concentration of activators bound to the origin of replication is neglected, so that the total (x_t), free (x_f) and BS bound (c) activator concentrations are related through

$$x_t = x_f + c. \tag{13}$$

The number of activators that can bind to each BS locus is denoted by N and the maximal concentration of bound activators is $N \cdot y$. Furthermore, we assume for simplicity that the free and bound activator concentrations are quickly equilibrated:

$$x_f \cdot (N \cdot y - c) = K_D \cdot c \Leftrightarrow x_f \cdot (N \cdot y - (x_t - x_f)) = K_D \cdot (x_t - x_f), \tag{14}$$

$$x_t - x_f = \frac{N \cdot y \cdot x_f}{K_D + x_f}. \tag{15}$$

Rearrangement in (15) gives

Table 1 Definitions and values of used parameters.

Definition of model parameter	Value
$k_{I\max}$ = constant of maximal initiation rate of replication	$0.1\ \mathrm{s}^{-1}$
K_I = dissociation constant of the activator binding to the origin	$8 \cdot 10^{-7}\ \mathrm{M}$
n_I = co-operativity factor for the activator binding to the origin	4
μ = rate constant for the cell growth rate	$1.9 \cdot 10^{-4}\ \mathrm{s}^{-1}$
k_s = rate constant of maximal synthesis rate of the activator	$10^6\ \mathrm{M}$
k_c = rate constant of conversion of (active) activator into inhibitor (inactive activator)	$9.8\ \mathrm{s}^{-1}$
K_a = dissociation constant of the inhibitor binding	$2 \cdot 10^{-6}\ \mathrm{M}$
n_a = co-operativity factor for inhibitor binding	4
k_d = degradation constant of inhibitors	$10\ \mathrm{s}^{-1}$
N = number of activators that can bind to the binding site (BS) outside the origin	500
K_D = dissociation constant for activator binding to the BS	$3.8 \cdot 10^{-9}\ \mathrm{M}$
M_{tot} = number of plasmid states in the synchronous model	100
k_R = rate constant of plasmid state change in the synchronous model	$1/(15 \cdot 60/M_{\mathrm{tot}})$

The given values of $k_{I\max}$, K_I, n_I, k_s, k_c, k_d, N and K_D in Table 1 correspond to $|A_{r_I,y}| = 71.1$. Intervals of co-varying constants for different values of $|A_{r_I,y}|$ are given in the figure legends. The value of k_s, k_c and k_d were adjusted to keep the steady state concentration of the plasmid unchanged and to always keep the inhibitor concentration proportional to the activator and plasmid concentration outside steady state.

$$x_t = x_f \cdot \left(1 + \frac{N \cdot y}{K_D + x_f}\right). \tag{16}$$

The parameters of the model (Table 1) were chosen so that proportionality between plasmid, activator and inhibitor concentrations was preserved.

2.2 Calculation of the Plasmid Adjustment Rate

By linearization of (10), the differential equation for the plasmid concentration, and by noticing that $dy/dt = d\Delta y/dt$, the rate of adjustment of the plasmid concentration back to steady state following a small deviation, Δy, can be studied. The linearized differential equation for a small deviation in plasmid concentration is

$$\frac{d\Delta y}{dt} = \left(k_I \cdot r_I \left(\overline{x_f} + \Delta x_f\right) - \mu\right) \cdot \left(\overline{y} + \Delta y\right). \tag{17}$$

For small Δx_f, r_I can be approximated by a first-order Taylor expansion giving

$$r_I\left(\overline{x_f} + \Delta x_f\right) \approx r_I\left(\overline{x_f}\right) + \frac{dr_I\left(\overline{x_f}\right)}{dx_f} \cdot \Delta x_f = \overline{r_I} + \frac{dr_I}{dx_f} \cdot \Delta x_f. \tag{18}$$

Inserting (18) into (17) generates

$$\frac{d\Delta y}{dt} = \left(k_I \cdot \left(\overline{r_I} + \frac{dr_I}{dx_f} \cdot \Delta x_f \right) - \mu \right) \cdot (\overline{y} + \Delta y). \tag{19}$$

From the definition of the sensitivity amplification factor follows that

$$\frac{d\Delta y}{dt} = \left(k_I \cdot \left(\overline{r_I} + \overline{A}_{r_I,x_f} \cdot \frac{\overline{r_I}}{\overline{x_f}} \cdot \Delta x_f \right) - \mu \right) \cdot (\overline{y} + \Delta y) \tag{20}$$

and

$$\frac{d\Delta y}{dt} = \left(k_I \cdot \overline{r_I} + k_I \cdot \overline{A}_{r_I,x_f} \cdot \frac{\overline{r_I}}{\overline{x_f}} \cdot \overline{A}_{x_f,y} \cdot \frac{\overline{x_f}}{\overline{y}} \cdot \Delta y - \mu \right) \cdot (\overline{y} + \Delta y). \tag{21}$$

Since at steady state $k_I \cdot \overline{r_I} = \mu$ (10) and by studying only small deviations from steady state so that $\Delta x_f \cdot (\overline{y} + \Delta y) \approx \Delta x_f \cdot \overline{y}$ give

$$\frac{d\Delta y}{dt} = \mu \cdot \overline{A}_{r_I,x_f} \cdot \overline{A}_{x_f,y} \cdot \Delta y = \mu \cdot \overline{A}_{r_I,y} \cdot \Delta y = -\mu \cdot \left| \overline{A}_{r_I,y} \right| \cdot \Delta y \tag{22}$$

and

$$\Delta y(t) = \Delta y(0) \cdot e^{-\mu \cdot \left| \overline{A}_{r_I,y} \right| \cdot t}. \tag{23}$$

Thus, the amplification factor $\left| \overline{A}_{r_I,y} \right|$ governs the rate by which the plasmid concentration adjusts back to steady state following a deviation in plasmid concentration and is therefore central for the quality of replication control. This has been recognized previously [7, 34] but not explicitly expressed in terms of $\left| \overline{A}_{r_I,y} \right|$.

2.3 Calculation of Sensitivity Amplification Factor

To study how sensitively the replication rate per plasmid responds to changes in the plasmid concentration, we calculated the sensitivity amplification factor at steady state as

$$\overline{A}_{r_I,y} = \left(\frac{dr_I}{dy} \right)_{y=\overline{y}} \cdot \frac{\overline{y}}{\overline{r_I}}. \tag{24}$$

$A_{\overline{r}_I,\overline{y}}$ can be obtained from the factorization

$$\overline{A}_{r_I,y} = \overline{A}_{r_I,x_f} \cdot \overline{A}_{x_f,y} = \left(\frac{dr_I}{dx_f} \right)_{x_f=\overline{x}_f} \cdot \frac{\overline{x_f}}{\overline{r_I}} \cdot \left(\frac{dx_f}{dy} \right)_{y=\overline{y}} \cdot \frac{\overline{y}}{\overline{x_f}}. \tag{25}$$

The first factor, $\overline{A}_{r_I,x_f}$, describing how sensitively the degree of activation of per plasmid origin responds to changes in the free concentration of the activator, is directly calculated as

$$\overline{A}_{r_I,x_f} = n_I \cdot (1 - \overline{r_I}). \tag{26}$$

To find the second factor, $\overline{A}_{x_f,y}$, describing how sensitively the free activator concentration responds to changes in the plasmid concentration, we studied the differential of (14), i.e.

$$dx_f \cdot \left(N \cdot \overline{y} - \left(\overline{x_t} - \overline{x_f}\right)\right) + \overline{x_f} \cdot \left(N \cdot dy - \left(dx_t - dx_f\right)\right) = K_D \cdot \left(dx_t - dx_f\right). \quad (27)$$

Rewriting and rearranging the terms give

$$\frac{dy}{\overline{y}} \cdot \left(N \cdot \overline{y} \cdot \overline{x_f}\right) - \frac{dx_t}{\overline{x_t}} \cdot \overline{x_t} \cdot \left(\overline{x_f} + K_D\right) = -\frac{dx_f}{\overline{x_f}} \cdot \overline{x_f} \cdot \left(N \cdot \overline{y} + K_D + \overline{x_f} - \left(\overline{x_t} - \overline{x_f}\right)\right). \quad (28)$$

A second relation between dx_t and dy can be found by studying the steady state differential of (11), i.e.

$$k_s \cdot dr_s \cdot \overline{y} + k_s \cdot \overline{r_s} \cdot dy = (k_c + \mu) \cdot dx_t \Leftrightarrow$$

$$k_s \cdot \frac{dr_s}{dx_t} \cdot dx_t \cdot \overline{y} + k_s \cdot \overline{r_s} \cdot \frac{dy}{\overline{y}} \cdot \overline{y} = (k_c + \mu) \cdot \frac{dx_t}{\overline{x_t}} \cdot \overline{x_t}. \quad (29)$$

By the steady state relation $\overline{x_d} = k_c \cdot \overline{x_t}/(k_d + \mu)$ from (12), the degree of activation of activator gene expression $\overline{r_s}$ can be described as a function of the total activator concentration $\overline{x_t}$. By noticing that

$$\overline{A}_{r_s,x_t} = \frac{dr_s}{dx_t} \cdot \frac{\overline{x_t}}{\overline{r_s}} = -n_a \cdot (1 - \overline{r_s}) \Leftrightarrow \frac{dr_s}{dx_t} = -n_a \cdot (1 - \overline{r_s}) \cdot \frac{\overline{r_s}}{\overline{x_t}} \quad (30)$$

equation (29) can be rewritten as

$$\frac{k_s \cdot \overline{r_s} \cdot \overline{y}}{(k_c + \mu) \cdot \overline{x_t}} \cdot \left(\frac{dy}{\overline{y}} - \frac{dx_t}{\overline{x_t}} \cdot n_a \cdot (1 - \overline{r_s})\right) = \frac{dx_t}{\overline{x_t}}. \quad (31)$$

From (11) it follows that $(k_s \cdot \overline{r_s} \cdot \overline{y})/((k_c + \mu) \cdot \overline{x_t}) = 1$ at steady state and (31) can be rewritten as

$$\frac{dx_t}{\overline{x_t}} = \frac{dy}{\overline{y}} \cdot \frac{1}{1 + n_a \cdot (1 - \overline{r_s})}. \quad (32)$$

Inserting (32) and (15) into (28) gives

$$\frac{dy}{\overline{y}} \cdot \left(\left(N \cdot \overline{y} \cdot \overline{x_f}\right) - \frac{1}{1 + n_a \cdot (1 - \overline{r_s})} \cdot \overline{x_f} \cdot \left(1 + \frac{N \cdot \overline{y}}{K_D + \overline{x_f}}\right) \cdot \left(\overline{x_f} + K_D\right)\right)$$

$$= -\frac{dx_f}{\overline{x_f}} \cdot \overline{x_f} \cdot \left(N \cdot \overline{y} + K_D + \overline{x_f} - \frac{N \cdot \overline{y} \cdot \overline{x_f}}{K_D + \overline{x_f}}\right). \quad (33)$$

Simplifications and rearrangements of factors finally give

$$\overline{A}_{x_f,y} = \frac{dx_f}{dy} \cdot \frac{\overline{y}}{\overline{x_f}} = -\left(1 - \frac{1}{1 + n_a \cdot (1 - \overline{r_s})} \cdot \left(1 + \frac{K_D + \overline{x_f}}{N \cdot \overline{y}}\right)\right) \cdot \frac{1}{\left(\frac{K_D + \overline{x_f}}{N \cdot \overline{y}} + \frac{K_D}{K_D + \overline{x_f}}\right)}. \quad (34)$$

The complete expression of the sensitivity amplification factor becomes

$$\overline{A}_{r_I,y} = -n_I \cdot (1 - \overline{r_I}) \cdot \left(1 - \frac{1}{1 + n_a \cdot (1 - \overline{r_s})} \cdot \left(1 + \frac{K_D + \overline{x_f}}{N \cdot \overline{y}}\right)\right) \cdot \frac{1}{\left(\frac{K_D + \overline{x_f}}{N \cdot \overline{y}} + \frac{K_D}{K_D + \overline{x_f}}\right)} \cdot \tag{35}$$

2.4 Analysis of the Sensitivity Amplification Factor

The numerical value of the sensitivity amplification $\left|\overline{A}_{r_I,y}\right|$ has three distinct factors, i.e.

$$f_1 = n_I \cdot (1 - \overline{r_I}), \tag{36}$$

$$f_2 = \frac{1}{\left(\frac{K_D + \overline{x_f}}{N \cdot \overline{y}} + \frac{K_D}{K_D + \overline{x_f}}\right)}, \tag{37}$$

and

$$f_3 = \left(1 - \frac{1}{1 + n_a \cdot (1 - \overline{r_s})} \cdot \left(1 + \frac{K_D + \overline{x_f}}{N \cdot \overline{y}}\right)\right). \tag{38}$$

The value of the amplification factor at steady state mainly varies by f_1 and f_2, while f_3 approximately equals $n_a/(n_a+1)$ for small values of $\overline{r_s}$ and for small values of $\left(K_D + \overline{x_f}\right)/(N \cdot \overline{y}) = \overline{x_f}/\overline{c}$ (15). Note that the value of factor f_3 ($f_3 < 1$) is a consequence of an imperfect regulation of the synthesis rate of the activator. When instead the total concentration of the activator is a true constant, the value of factor f_3 becomes 1. This is seen by studying the differential of (14) when x_t is a constant. A constant total concentration of the activator corresponds to an infinitely sensitive regulation of activator synthesis and an infinitely high value of the co-operativity factor n_a ($f_3 \underset{n_a \to \infty}{\to} 1$, (38)). Then, only f_1 and f_2 govern the value of $\left|\overline{A}_{r_I,y}\right|$.

The value of factor f_1 increases with increasing value of the co-operativity factor n_I of activator binding to the origin and a decreasing value of $\overline{r_I}$, the degree of activation per origin at steady state. The value of $\overline{r_I}$ at steady state is a function of $k_{I\max}$, the maximal rate of initiation of replication. A high co-operativity factor is a well-known way of increasing the absolute value of a sensitivity amplification factor (Koshland et al. 1982) and a low $\overline{r_I}$ (a high $k_{I\max}$) maximizes the value of $\left|\overline{A}_{r_I,y}\right|$ within the limit set by the regulatory mechanism as previously shown by Paulsson and Ehrenberg [7]. Figure 2a shows that $\left|\overline{A}_{r_I,y}\right|$ increases approximately by the same factor as n_I increases. To be able to accurately compare one value of $\left|\overline{A}_{r_I,y}\right|$ to another the plasmid concentration ($\overline{y}$) must stay unaltered. Therefore, when f_1 was changed by changing n_I in the model, $k_{I\max}$ was co-varied to keep the replication rate per plasmid and thereby the plasmid concentration at steady state remained unchanged.

Factor f_2 represents a novel factor of sensitivity amplification, originating from this type of replication control circuit with one dominating binding site that titrates

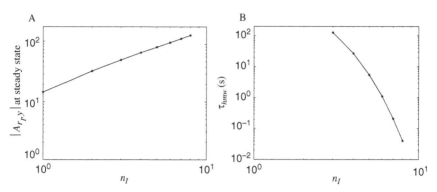

Fig. 2 (**A**) The absolute value of the sensitivity amplification factor $\left|\overline{A}_{r_I,y}\right|$ as a function of the co-operativity factor n_I for activator binding to the origin in the asynchronous model. The rate constant of maximal initiation of replication $k_{I\max}$ was co-varied between 1.1×10^{-3} and 51.7 s^{-1} for $n_I = 1$–8. (**B**) The width, τ_{hmw}, of the replication rate, $k_{I\max} \cdot r_I \cdot y_r$, at half its maximal value at initiation as a function of n_I in the synchronous model. $k_{I\max}$ co-varied between $2.1 \cdot 10^{-2}$ and 51.7 s^{-1}. For low values of n_I ($n_I = 1$ and 2) the model is either asynchronous or not completely synchronous and no well-defined τ_{hmw} exists

activator molecules with a higher binding affinity than activator binding to the origin. Factor f_2 attains the maximal value

$$f_2 = \frac{\sqrt{N \cdot \overline{y}}}{2 \cdot \sqrt{K_D}} \tag{39}$$

for the variable $K_D + \overline{x_f}$ when it holds that

$$K_D + \overline{x_f} = \sqrt{K_D \cdot N \cdot \overline{y}}. \tag{40}$$

When (40) holds and when $K_D \to 0$, the ratio between the maximal concentration of activators that can bind to BS and the free activator concentration at steady state, $N \cdot \overline{y}/\overline{x_f}$, and the ratio between the free activator concentration at steady state and the dissociation constant for activator binding to BS, $\overline{x_f}/K_D$, becomes the $\sqrt{N \cdot \overline{y}}/\sqrt{K_D}$ ratio appearing in the expression of the maximal value of f_2 (39)

$$\frac{N \cdot \overline{y}}{\overline{x_f}} \approx \frac{\overline{x_f}}{K_D} \approx \frac{\sqrt{N \cdot \overline{y}}}{\sqrt{K_D}}. \tag{41}$$

The free activator concentration at steady state, $\overline{x_f}$, is determined by the dissociation constant of activator binding to the origin K_I. A high value of f_2 and of $\left|\overline{A}_{r_I,y}\right|$ therefore correspond to high N/K_I and K_I/K_D ratios. We studied the value of $\left|\overline{A}_{r_I,y}\right|$ by co-varying $\overline{x_f}$ and K_D, or $\overline{x_f}$ and N, or N and K_D (40) so that the value of factor f_2 was maximal (39). The values of k_s, k_c and k_d were changed to keep the total activator concentration and the plasmid concentration unchanged at steady state (14). Figure 3a shows how $\left|\overline{A}_{r_I,y}\right|$ increases approximately by the same factor as

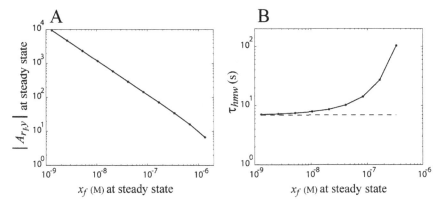

Fig. 3 (**A**) The absolute value of the sensitivity amplification factor $\left|\overline{A}_{r_I,y}\right|$ as a function of the free concentration of the activator $\overline{x_f}$ at steady state in the asynchronous model. The dissociation constant for binding to the binding site outside the origin K_D was co-varied between $3.9 \cdot 10^{-7}$ and $2.2 \cdot 10^{-13}$ M for $\overline{x_f}$ ranging between $1.3 \cdot 10^{-6}$ and $1.3 \cdot 10^{-9}$ M. (**B**) The width, τ_{hmw}, of the replication rate, $k_{I\max} \cdot r_I \cdot y_r$, at half its maximal value at initiation in the synchronous model as a function of $\overline{x_f}$ at steady state in the asynchronous model. $\overline{x_f}$ is varied by changing K_I, the dissociation constant of activator binding to the origin. K_D co-varied between $1.6 \cdot 10^{-8}$ and $2.2 \cdot 10^{-3}$ M. For high values of $\overline{x_f}$ ($\overline{x_f} > \approx 4 \cdot 10^{-7}$ M) the model is either asynchronous or not completely synchronous and no well-defined τ_{hmw} exists. The *broken line* marks the lower limit of τ_{hmw}

$\overline{x_f}$ decreases when K_D is co-varied. In Fig. 4a, N is co-varied with $\overline{x_f}$ and $\left|\overline{A}_{r_I,y}\right|$ increases by the square root of the factor N (39). Plotting $\left|\overline{A}_{r_I,y}\right|$ as a function of N when K_D is co-varied generates the mirror image of Fig. 3a (not shown).

2.5 Synchronous Plasmid Model

Initiation synchrony can be created by introducing a time-delay between replication of the origin and the replication of BS, along with different plasmid states, where only one state is capable of initiating replication. In the synchronous model a plasmid can be present in 100 different states ($i = 1, 2, 3, \ldots, 99, r$) with concentrations ($y_1, y_2, \ldots, y_{99}, y_r$), which for instance could represent different positions of the replication fork during replication:

$$\frac{dy_1}{dt} = 2 \cdot k_{I\max} \cdot r_I \cdot y_r - (k_R + \mu) \cdot y_1, \tag{42}$$

$$\frac{dy_i}{dt} = k_R \cdot y_{i-1} - (k_R + \mu) \cdot y_i, \quad i = 2, \ldots, 99 \tag{43}$$

$$\frac{dy_r}{dt} = k_R \cdot y_{99} - (k_{I\max} \cdot r_I + \mu) \cdot y_r. \tag{44}$$

Plasmids in states $1, 2, \ldots, 99$ are refractory to initiation of replication, which can occur only for plasmids in state r. Initiation of replication from the latter state occurs

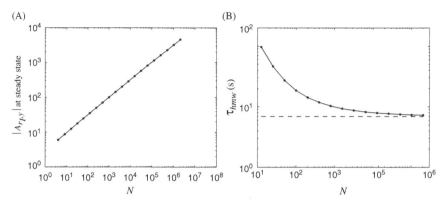

Fig. 4 (**A**) The absolute value of the sensitivity amplification factor $|\overline{A}_{r_l,y}|$ as a function of the number of activators N that can bind to the binding site outside the origin in the asynchronous model. The free concentration of the activator $\overline{x_f}$ at steady state varied between $1.2 \cdot 10^{-8}$ and $1.1 \cdot 10^{-5}$ M for N between 4 and $2.048 \cdot 10^6$. (**B**) The width, τ_{hmw}, of the replication rate, $k_{I\max} \cdot r_l \cdot y_r$, at half its maximal value at initiation in the synchronous model as a function of N. K_D co-varied between $8.2 \cdot 10^{-8}$ and $1.1 \cdot 10^{-5}$ M. For low values of N ($N < 125$) the model is either asynchronous or not completely synchronous and no well-defined τ_{hmw} exists. The *broken line* marks the lower limit of τ_{hmw}

with the compounded rate constant $k_{I\max} \cdot r_l$, as in (10). After initiation, two plasmids in state 1 are created. They leave state 1 and move through all successive states by the replication rate constant k_R until they reach state r, from where a new replication cycle starts. All concentrations are reduced by volume increase, determined by the growth rate μ as in (10–12). The replication of locus BS for activator binding is placed in plasmid state 50 and, hence, the increase in its concentration [BS] after initiation of replication will be delayed by a well-defined time defined by the movement of a truncated Poisson process from state 1 to state 50:

$$\frac{d\,[\mathrm{BS}]}{dt} = 1/2 \cdot k_R \cdot y_{50} - \mu \cdot [\mathrm{BS}]. \tag{45}$$

Besides replacing the plasmid concentration, y, by the sum of the concentration of plasmids in all the different states in (11), and replacing y by [BS] in (14), all other equations are the same and all rate constants are the same as in the asynchronous model. As before, the model parameters were chosen to assure approximate proportionality between the plasmid, activator and inhibitor concentrations (Table 1).

2.6 High Sensitivity Amplification is Required for Initiation Synchrony

When the absolute value of the sensitivity amplification factor $|\overline{A}_{r_l,y}|$ increases, as defined in the asynchronous model, the model defined by (42–44) switches from asynchronous to synchronous initiation of replication with a visible cycle time

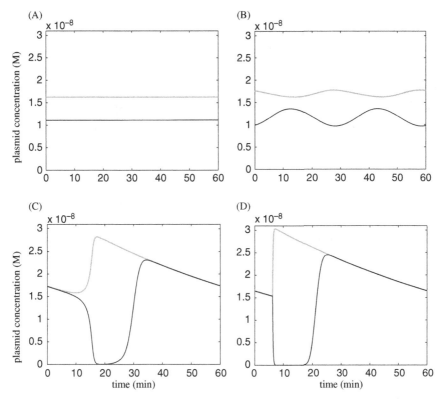

Fig. 5 Variations in the total (the sum of all plasmid states, *grey line*) and initiation proficient (y_r, *black line*) concentrations of plasmids during one cell generation (one cycle) in the synchronous model. The corresponding sensitivity amplification factor $\left|\bar{A}_{r_I,y}\right|$ at steady state in the asynchronous model for the same set of parameter values is in (**A**) $\left|\bar{A}_{r_I,y}\right| = 6.7$, in (**B**) $\left|\bar{A}_{r_I,y}\right| = 16.1$, in (**C**) $\left|\bar{A}_{r_I,y}\right| = 34.5$ and in (**D**) $\left|\bar{A}_{r_I,y}\right| = 2286.8$. The values of $\left|\bar{A}_{r_I,y}\right|$ were calculated by co-varying $\overline{x_f}$ (K_I) and K_D

(Fig. 5a–c). Further increase of $\left|\bar{A}_{r_I,y}\right|$ results in a more sharply defined replication time (compare Fig. 5c with 5d and Fig. 6, top figures). To quantify this latter effect, we studied the width, τ_{hmw}, of the replication rate, $k_{I\max} \cdot r_I \cdot y_r$, (Fig. 6 (middle figures)) at half its maximal value, as a function of $\left|\bar{A}_{r_I,y}\right|$ (Fig. 2–4). Comparing (a) and (b) in Fig. 2–4 reveals that an increasing $\left|\bar{A}_{r_I,y}\right|$ corresponds to a decreasing τ_{hmw}. The width, τ_{hmw}, at initiation is determined by how quickly the replication rate increases and decreases at initiation. How quickly the replication rate increases at initiation is governed by the rate r_I, the degree of activation per origin, increases. How quickly the replication rate decreases is a function of $k_{I\max}$, the maximal rate of initiation, since μ, the rate of plasmid dilution, is much slower. By increasing the value of n_I, the factor of co-operativity for activator binding to the origin r_I is turned on within a narrower interval of x_f, the free activator concentration, and r_I starts acting like a switch that is momentarily turned on at a certain x_f. Thus, when n_I and $k_{I\max}$ simultaneously increase (increasing $\left|\bar{A}_{r_I,y}\right|$ by increasing factor f_1 (36)

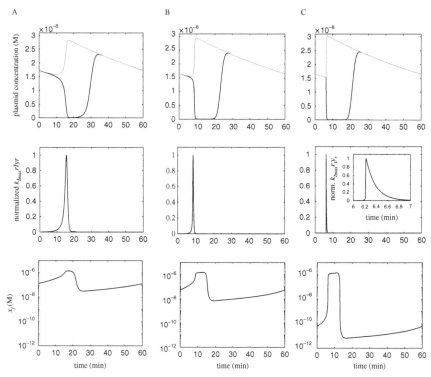

Fig. 6 The plasmid concentration (*top*), normalized replication initiation rate ($k_{Imax} \cdot r_I \cdot y_r$) (*middle*) and the free concentration of the activator (x_f) (*bottom*) during one cell generation (one cycle) in the synchronous model. The total plasmid concentration (the sum of all plasmid states) is shown with a *grey solid line* and the concentration of initiation proficient plasmids (y_r) is shown with a *black solid line*. Initiation synchrony increases from the left figures to the right figures by varying K_I (the dissociation constant for activator binding to the origin) and K_D (the dissociation constant for activator binding to the binding site outside the origin). The width of the peak of the normalized replication initiation rate at 0.5, τ_{hmw}, (*middle figures*) is in (**A**) 103.3 s, in (**B**) 27.1 s and in (**C**) 7.5 s. The corresponding sensitivity amplification factor $\left| \bar{A}_{r_I,y} \right|$ at steady state in the asynchronous model for the same set of parameter values is in (**A**) $\left| \bar{A}_{r_I,y} \right| = 34.5$, in (**B**) $\left| \bar{A}_{r_I,y} \right| = 71.1$ and in (**C**) $\left| \bar{A}_{r_I,y} \right| = 2286.8$. The inserted middle figure in (**C**) is a blow up of the peak of the replication initiation rate and shows that at a high initiation synchrony close to the lower limit of τ_{hmw} (here $\tau_{hmw} \approx \ln 2/0.1 \approx 6.93$, Fig. 3b), the time τ_{hmw} is determined by the exponential decrease of y_r at rate k_{Imax}

in the asynchronous model) the width τ_{hmw} decreases in the synchronous model. For a fixed value of n_I and k_{Imax}, r_I is turned on momentarily when the ratios N/K_I and K_I/K_D simultaneously approaches infinity. The product $N \cdot y$ is the maximal concentration of activators that can bind to BS, the binding site outside the origin, and K_I determines at what free concentrations of the activator replication initiates. For an approximately unchanged value of the plasmid concentration and of the total activator concentration, an increasing N/K_I ratio means that the concentration of free activator required to initiate replication decreases in relation to the concentration of activators bound to the binding site (BS) outside the origin, which decreases the time, following saturation of BS, for free activators to reach the concentration at which they trigger replication. In addition, a high K_I/K_D ratio separates the activator binding to BS from the activator binding to the origin, i.e. no plasmids initiate replication (no activators bind to the origin) until BS is saturated by activators. Without interference by activator binding to BS, the free activator concentration increases faster and results in a high and well-defined peak of the free concentration of activators at initiation (Fig. 6, bottom figures). The degree of origin activation, r_I, reaches its maximal value ($r_I = 1$) so that initiation of replication takes place with its maximal rate, $k_{I\text{max}}$. Thus, when the N/K_I and K_I/K_D ratios simultaneously increase ((41), $\overline{x_f}$ is defined by K_I), corresponding to increasing values of $\left|\overline{A}_{r_I,y}\right|$ by increasing factor f_2 (39) in the asynchronous model, the width τ_{hmw} decreases in the synchronous model. The value of τ_{hmw} is uniquely defined by the N/K_I and K_I/K_D ratios and is independent of the absolute values of N, K_D and K_I. Therefore, there is a one-to-one relation between $\left|\overline{A}_{r_I,y}\right|$ and τ_{hmw} (Fig. 7).

When r_I increases momentarily, τ_{hmw} is determined by the rate it takes the plasmids in state y_r to initiate at maximal speed

$$\frac{\mathrm{d}y_r(t)}{\mathrm{d}t} = -k_{I\text{max}} \cdot y_r \quad \text{and} \quad y_r(t) = y_{r_{\text{max}}} \cdot \mathrm{e}^{-k_{I\text{max}} \cdot t}, \tag{46}$$

where $y_{r_{\text{max}}}$ is the maximal value of the concentration of plasmids capable of initiating replication during a cycle. In the limit when r_I is maximized momentarily ($t_{\text{max}r_I} \to 0$) along with a fixed finite value of $k_{I\text{max}}$, a lower limit of τ_{hmw} is found

$$\lim_{t_{\text{max}r_I} \to 0} \tau_{\text{hmw}} \approx \frac{\ln 2}{k_{I\text{max}}}. \tag{47}$$

which is why τ_{hmw} approaches a lower limit in Fig. 3b and 4b where $k_{I\text{max}}$ is fixed but not in Fig. 2b where $k_{I\text{max}}$ is monotonously increasing.

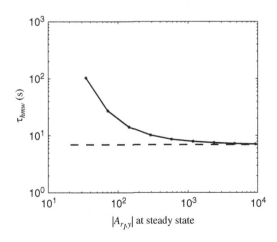

Fig. 7 The width, τ_{hmw}, of the replication rate, $k_{I\text{max}} \cdot r_I \cdot y_r$, at half its maximal value at initiation in the synchronous model as a function of the sensitivity amplification factor $\left|\overline{A}_{r_I,y}\right|$ at steady state in the asynchronous model when the factor f_2 is varied as described in the text. The *broken line* marks the lower limit of τ_{hmw}

3 Conclusions

3.1 Relevance of the Analysis for Chromosomal Replication Control

We have presented a novel mechanism for the control of initiation of replication that resolves the auto-regulation sequestration paradox, i.e. how a molecule sequestered by binding to DNA may at the same time accurately regulate its own synthesis [6]. The novel control circuit contains an activator molecule and an activator titrating binding site outside the origin. The activator binds to the origin and triggers initiation of replication after saturation of the binding site outside the origin. Activators are converted into unstable inhibitors by a high rate and the inhibitors regulate activator synthesis. The free activator concentration does not represent the total activator concentration due to sequestration of activators to the binding site. Therefore, the free activator concentration cannot regulate activator synthesis accurately – the so-called auto-regulation sequestration paradox. However, the rapid conversion of activators into inhibitors creates proportionality between activator and inhibitor concentration and allows the inhibitor concentration to accurately control the total concentration of the activator which elegantly solves the auto-regulation paradox.

We analysed the novel mechanism for replication control in the context of an asynchronous and a synchronous plasmid model. The analysis revealed that in the asynchronous model the central sensitivity amplification factor, $\left|\overline{A}_{r_I,y}\right|$, that measures how sensitively the replication frequency per plasmid ($k_{I\max} \cdot \overline{r_I}$) responds to changes in plasmid concentration ($\overline{y}$) at steady state can be divided into three factors. The first factor, factor f_1 (36) is the product of the co-operativity factor for activator binding to the origin, n_I, and the degree of repression per origin, $1 - \overline{r_I}$. The value of $\overline{r_I}$ is a function of $k_{I\max}$, the maximal rate of initiation of replication. Both a high co-operativity factor and a high degree of origin repression (a high value of $k_{I\max}$) have previously been recognized to increase sensitivity [7, 24, 34]. We also found a new and to our knowledge previously not described factor which increases sensitivity, characteristic for this type of control circuit, with one dominating binding site (BS) outside the origin which binds activator molecules with higher affinity than activator binding to the origin. The second factor f_2 (37) is a function of the dissociation constant of activator binding (K_D), the maximal concentration of activators that can bind to BS ($N \cdot \overline{y}$) and the free activator concentration ($\overline{x_f}$) at steady state. A high value of f_2 is a function of a high $N \cdot \overline{y}/K_D$ ratio and corresponds to high N/K_I and K_I/K_D ratios ((39) and (41), $\overline{x_f}$ is defined by K_I). The third factor, factor f_3 (38), is a degenerative factor. It results from the regulation of the activator synthesis rate with finite sensitivity. With a perfect regulation, resulting in a truly constant total activator concentration (x_t), f_3 attains its maximal value of 1 and only factors f_1 and f_2 determine the value of $\left|\overline{A}_{r_I,y}\right|$.

A high-sensitivity amplification factor, $\left|\overline{A}_{r_I,y}\right|$, in the asynchronous model corresponds to a high initiation synchrony, measured as the width, τ_{hmw}, of the replication rate, $k_{I\max} \cdot r_I \cdot y_r$, at half its maximal value, in the synchronous model. Thus, the same parameters that create a sensitive response in replication frequency to changes in plasmid concentration (a high $|\overline{A}_{r_I,y}|$) in the asynchronous model create initiation synchrony (a low τ_{hmw}) in the synchronous model (Figs. 2–4). A too low value of $|\overline{A}_{r_I,y}|$ corresponds to asynchronous initiations of replication also in the synchronous model (Fig. 5). The width, τ_{hmw}, decreases by the rate the degree of origin activation, r_I, is turned on and by the rate of maximal initiation of replication, k_{Imax}. A high co-operativity factor n_I switches on r_I to its maximal value within a narrow range of concentrations of free activator x_f. In addition, a high N/K_I ratio, at which the concentration of free activator required to initiate replication is low in relation to the concentration of activators bound to BS outside the origin, and a high K_I/K_D ratio, that separates activator binding to BS from activator binding to the origin, decreases the time for the free activator to reach the concentration that triggers initiation of replication. High N/K_I and K_I/K_D ratios correspond to a high factor f_2 in the asynchronous model (39 and 41), where $\overline{x_f}$ is determined by K_I. In the limit when r_I increases momentarily, the width τ_{hmw} is solely determined by the finite value of k_{Imax}, $\tau_{hmw} \approx \ln 2/k_{Imax}$ ((47), Fig. 6c (middle)).

The inspiration of the novel regulatory mechanism comes from the molecular set-up of the regulation of initiation of chromosome replication in the bacterium E. coli. The idea of presenting the novel mechanism of replication control in the context of plasmid control was to illustrate principles in a simple fashion, relevant for both types of replication control, plasmids as well as chromosomes, and to relate the analysis to the previous analysis of inhibitor-dilution regulated plasmids [7]. The novel regulatory mechanism behaves similarly for plasmids and the chromosome. In both cases, the dominating cluster of activator binding sites (BS) outside oriC regulates the oriC concentration, i.e. too high concentration of oriC (and BS) reduces the free activator concentration and vice versa.

The major difference between chromosomal and plasmid replication control not accounted for in the synchronous plasmid model is that the chromosomal gene expression and synthesis rate of an activator is not proportional to the origin concentration but instead expected to be approximately invariant at a given growth rate [27]. With a constant activator synthesis rate, independent of the origin concentration of the chromosome, a control circuit of replication initiation may solely consist of a constitutive, unregulated synthesis of an activator that initiates replication by binding to the origin [27]. If, on the other hand, the activator synthesis rate is proportional to the origin concentration as for plasmids, constitutive synthesis of an activator would result in inadequate replication control manifested as run-away replication. However, here activator synthesis is regulated, which makes the difference in constant versus proportional gene expression unimportant. Thus, the factors important for creating initiation synchrony of plasmid replication are also important for creating initiation synchrony of chromosomal replication. Translated into the terminology of E. coli, initiation synchrony is expected to be a function of both the values of the dissociation constants of DnaA-ATP binding to oriC (K_I) and to the datA locus (K_D) as well as co-operative binding of DnaA-ATP to the origin (n_I), and the rate replication is triggered once DnaA-ATP is bound (k_{Imax}). A consequence of an increasing initiation synchrony is that the datA locus, the dominating binding site, can be positioned closer to the origin without disrupting synchronous initiations of replication.

In the synchronous plasmid model the inhibitor (the inactive form of the activator) was continuously synthesized and actively degraded by a high rate. This scenario is distinct from what is known about DnaA-ADP, the inactive form of the activator of chromosomal replication in *E. coli*, where DnaA-ADP is stable and produced by the conversion of DNA-bound activators (activators in the active form), DnaA-ATP, as the DNA molecule becomes replicated, in particular the *datA* locus [11]. Still, conversion of an approximately fixed number of activators molecules in every round of replication into their inactive form also solves the auto-regulation sequestration paradox. With one dominating binding site, most of the DnaA-ATP molecules are expected to be DNA-bound at the *datA* locus and thus the number of converted DnaA-ATPs at the *datA* loci approximates the total number of DnaA-ATP molecules in the cell at the initiation of replication. The free concentration of DnaA-ADP in the cell will then reflect the flow of DnaA-ATP to DnaA-ADP, proportional to the total DnaA-ATP concentration and the dilution of DnaA-ADP by volume growth and the synthesis rate of active activators, DnaA-ATP, can be regulated adequately.

References

1. Messer W (2002) The bacterial replication initiator DnaA. DnaA and *oriC*, the bacterial mode to initiate DNA replication. FEMS Microbiol. Rev. 26:355–374
2. Summers DK (1996) The biology of plasmids. Blackwell Science, Oxford
3. Boye E, Løbner-Olesen A and Skarstad K (2000) Limiting DNA replication to once and only once. EMBO Rep. 1:479–483
4. Cooper S and Helmstetter CE (1968) Chromosome replication and the division cycle of *Escherichia coli* B/r. J. Mol. Biol. 31:519–540
5. Nordström K and Austin SJ (1993) Cell-cycle-specific initiation of replication. Mol. Microbiol. 10:457–463
6. Chattoraj DK, Mason RJ and Wickner SH (1988) Mini-P1 plasmid replication: The auto-regulation sequestration paradox. Cell 52:551–557
7. Paulsson J and Ehrenberg M (2001) Noise in a minimal regulatory network: Plasmid copy number control. Q. Rev. Biophys. 34:1–59
8. Skarstad K, Boye E and Steen HB (1986) Timing of initiation of chromosome replication in individual *Escherichia coli* cells. EMBO J. 5:1711–1717
9. Marsh RC and Worcel A (1977) A DNA fragment containing the origin of replication of the *E. coli* chromosome. Proc. Natl. Acad. Sci. USA 74:2720–2724
10. Kitagawa R, Mitsuki H, Okazaki T and Ogawa T (1996) A novel DnaA protein-binding site at 94.7 min on the *Escherichia coli* chromosome. Mol. Microbiol. 19:1137–1147
11. Katayama T, Kubota T, Kurokawa K, Crooke E and Sekimizu K (1998) The initiator function of DnaA protein is negatively regulated by the sliding clamp of the *E. coli* chromosomal replicase. Cell 94:61–71
12. Bochner BR and Ames BN (1982) Complete analysis of cellular nucleotides by two-dimensional thin layer chromatography. J. Biol. Chem. 257:9759–9769
13. Erzberger JP, Mott ML and Berger JM (2006) Structural basis for ATP-dependent DnaA assembly and replication-origin remodeling. Nat. Struct. Mol. Biol. 13:676–683

14. Campbell JL and Kleckner N (1990) *E. coli oriC* and the *dnaA* gene promoter are sequestered from *dam* methyltransferase following the passage of the chromosomal replication fork. Cell 62:967–979

15. Speck C, Weigel C and Messer W (1999) ATP- and ADP-DnaA protein, a molecular switch in gene regulation. EMBO J. 18:6169–6176

16. Dasgupta S and Løbner-Olesen A (2004) Host controlled plasmid replication: *Escherichia coli* minichromosomes. Plasmid 52:151–168

17. Polinsky B (1988) ColE1 replication control circuitry: Sense from antisense. Cell 55:929–932

18. Nordström K and Wagner EGH (1994) Kinetic aspects of control of plasmid replication by antisense RNA. Trends Biochem. Sci. 19:294–300

19. Das N, Valjavec-Gratian M, Basuray AN, Fekete RA, Papp PP, Paulsson J and Chattoraj DK (2005) Multiple homeostatic mechanisms in the control of P1 plasmid replication. Proc. Natl. Acad. Sci. USA 102:2856–2861

20. Uga H, Matsunaga F and Wada C (1999) Regulation of DNA replication by iterons: An interaction between the *ori2* and *incC* regions mediated by RepE-bound iterons inhibits DNA replication of mini-F plasmid in *Escherichia coli*. EMBO J. 18:3856–3867

21. Pal SK and Chattoraj DK (1988) P1 plasmid replication: Initiator sequestration is inadequate to explain control by initiator-binding sites. J. Bacteriol. 170:3554–3560

22. McEachern MJ, Bott MA, Tooker PA and Helinski DR (1989) Negative control of plasmid R6K replication: Possible role of intermolecular coupling of replication origins. Proc. Natl. Acad. Sci. USA 86:7942–7946

23. Savageau MA (1971) Parameter sensitivity as a criterion for evaluating and comparing the performance of biochemical systems. Nature 229:542–544

24. Koshland DE, Goldbeter A and Stock JB (1982) Amplification and adaptation in regulatory and sensory systems. Science 217:220–225

25. Jacob F, Brenner S and Cuzin F (1963) On the regulation of DNA replication in bacteria. Cold Spring Harb. Symp. Quant. Biol. 28:329–348

26. Pritchard RH, Barth PT and Collins J (1969) Control of DNA synthesis in bacteria. Symp. Soc. Gen. Microbiol. 19:263–297

27. Bremer H and Churchward G (1991) Control of cyclic chromosome replication in *Escherichia coli*. Microbiol. Rev. 55:459–475

28. Hansen FG, Christensen BB and Atlung T (1991) The initiator titration model: Computer simulation of chromosome and minichromosome control. Res. Microbiol. 142:161–167

29. Nordström K (2003) The replicon theory 40 years. Plasmid 49:269–280

30. Fantes PA, Grant WD, Pritchard RH, Sudbery PE and Wheals AE (1975) The regulation of cell size and the control of mitosis. J. Theor. Biol. 50:213–244

31. Donachie WD and Blakely GW (2003) Coupling the initiation of chromosome replication to cell size in *Escherichia coli*. Curr. Opin. Microbiol. 6:146–150

32. Kurokawa K, Nishida S, Emoto A, Sekimizu K and Katayama T (1999) Replication cycle-coordinated change of the adenine nucleotide-bound forms of DnaA protein in *Escherichia coli*. EMBO J. 18:6642–6652

33. Paulsson J (2004) Summing up the noise in gene networks. Nature 427:415–418

34. Paulsson J, Nordström K and Ehrenberg M (1998) Requirements for rapid plasmid ColE1 copy number adjustments: A mathematical model of inhibition modes and RNA turnover rates. Plasmid 39:215–234

35. Paulsson J and Ehrenberg M (1998) Trade-off between segregational stability and metabolic burden: a mathematical model of plasmid ColE1 replication control. J. Mol. Biol. 279:73–88

Activity-Dependent Model
for Neuronal Avalanches

L. de Arcangelis

Abstract Networks of living neurons represent one of the most fascinating systems of modern biology. If the physical and chemical mechanisms at the basis of the functioning of a single neuron are quite well understood, the collective behavior of a system of many neurons is an extremely intriguing subject. Crucial ingredient of this complex behavior is the plasticity property of the network, namely the capacity to adapt and evolve depending on the level of activity. This plastic ability is believed, nowadays, to be at the basis of learning and memory in real brains. This fundamental problem in neurobiology has recently shown a number of features in common to other complex systems. These features mainly concern the morphology of the network, namely the spatial organization of the established connections, and a novel kind of neuronal activity. Experimental data have, in fact, shown that electrical information propagates in a cortex slice via an avalanche mode. Both features have been found in other problems in the context of the physics of complex systems and successful models have been developed to describe their behavior. In this contribution, we apply a statistical mechanical model to describe the complex activity in a neuronal network. The network is chosen to have a number of connections in long range, as found for neurons in vitro. The model implements the main physiological properties of living neurons and is able to reproduce recent experimental results. The numerical power spectra for electrical activity reproduces also the power law behavior measured in an EEG of man resting with the eyes closed.

1 Introduction

Cortical networks exhibit diverse patters of activity, including oscillations, synchrony and waves. During neuronal activity, each neuron can receive inputs by thousands of other neurons and, when it reaches a threshold, redistributes this integrated activity back to the neuronal network. Recently, it has been shown that another mode

de L. Arcangelis
Department of Information Engineering and CNISM, Second University of Naples, 81031 Aversa (CE), Italy, dearcangelis@na.infn.it

de Arcangelis, L.: *Activity-Dependent Model for Neuronal Avalanches*. Lect. Notes Phys. **752**, 215–230 (2008)
DOI 10.1007/978-3-540-78765-5_10

of activity is neuronal avalanches, with a dynamics similar to self-organized criticality (SOC) [1, 2, 3, 4], observed in organotypic cultures from coronal slices of rat cortex [5] where neuronal avalanches are stable for many hours [6]. The term SOC usually refers to a mechanism of slow energy accumulation and fast energy redistribution driving the system toward a critical state, where the distribution of avalanche sizes is a power law obtained without fine tuning: no tunable parameter is present in the model. The simplicity of the mechanism at the basis of SOC has suggested that many physical and biological phenomena characterized by power laws in the size distribution represent natural realizations of the SOC idea. For instance, SOC has been proposed to model earthquakes [7, 8], the evolution of biological systems [9], solar flare occurrence [10], fluctuations in confined plasma [11] snow avalanches [12], and rain fall [13].

In order to monitor neural activities, different time series are usually analyzed through power spectra and generally power-law decay is observed. A large number of time series analyses have been performed on medical data that are directly or indirectly related to brain activity. Prominent examples are EEG data which are used by neurologists to discern sleep phases, diagnose epilepsy, and other seizure disorders as well as brain damage and disease [14, 15]. However, the interpretation of physiological mechanisms at the basis of EEG measurements is still controversial. Another example of a physiological function which can be monitored by time series analysis is the human gait which is controlled by the brain [16]. For all these time series the power spectrum, i.e. the square of the amplitude of the Fourier transformation double logarithmically plotted against frequency, generally features a power law at least over one or two orders of magnitude with exponents between 1 and 0.7. Moreover, experimental results show that neurotransmitter secretion rate exhibits fluctuations with time power law behavior [17] and power laws are observed in fluctuations of extended excitable systems driven by stochastic fluctuations [18].

Here we discuss a model based on SOC ideas, taking into account synaptic plasticity in a neural network [19]. With this model we analyze the time signal for electrical activity and compare the power spectra with EEG data. Plasticity is one of the most astonishing properties of the brain, occurring mostly during development and learning [20, 21, 22], and can be defined as the ability to modify the structural and functional properties of synapses. In the mammalian central nervous system, the refinement of neuronal connections is thought to occur during "critical periods" of early postnatal life, when circuits are particularly susceptible to electrical activity triggered by external sensory inputs. Modifications in the strength of synapses are thought to underlie memory and learning. Among the postulated mechanisms of synaptic plasticity, the activity dependent Hebbian plasticity constitutes the most fully developed and influential model of how information is stored in neural circuits [23, 24, 25, 26]. A large variety of models for brain activity has been proposed, based, for instance, on the convolution of oscillators [27] or stochastic waiting times [28]. They are essentially abstract representations on a mesoscopic scale, but none of them is based on the behavior of a neural network itself. In order to get real insights on the relation between time series and the microscopic, i.e. cellular, interactions inside a neural network, it is necessary to identify the basic ingredients of the brain activity possibly responsible for characteristic scale-free behavior observed through the spectrum power law. This insight is the basis for any further understanding of the diverse additional features that are observed and interpreted by practitioners that analyze these time series for diagnosis. Therefore the formulation of a brain model that yields the correct power spectrum is of crucial importance for any further progress in the understanding of the living brain.

2 Neurons, Synapses, and Hebbian Plasticity

In the early twentieth century, the first generation of "modern" neurobiologists had difficulty in interpreting the cellular nature of the nervous system, since this exhibits an extraordinary complexity [29] with respect to other tissues. The Spanish neuroanatomist Ramón y Cajal first argued that nerve cells are discrete entities, which communicate with each other by means of specialized contacts, successively named "synapses." Interestingly, Cajal came to his conclusions by light microscopic examinations of nervous tissue stained with silver salts, according to a method introduced by Camillo Golgi, an Italian neuropathologist. Golgi was instead in favour of the "reticular theory," viewing the nervous system as a continuous nerve cell network or "reticulum." Surprisingly, Golgi and Cajal were jointly recognized by the award of the Nobel Prize for Medicine in 1906, even if they were supporting opposite interpretations for the organization of the nervous system. The advent of electron microscopy in the 1950s finally established that nerve cells are indeed functionally independent units, as first understood by Cajal.

It is now widely accepted that cells in the nervous system can be divided into two categories: neurons and a variety of supporting cells, as neurological cells. Neurons are specialized for electrical signaling over long distances, whereas supporting cells are not capable of electrical signaling, despite having an important role in signal transmission. The human brain is estimated to contain 100 billion neurons and several times as many supporting cells. The interconnection of nerve cells via synapses forms an intricate network, which is the foundation on which sensory processes, perception, and behavior are built (Fig. 1).

Each neuron is made of a cell body from which arise one axon and an eventually elaborate arborization of dendrites. These receive inputs from other neurons and represent the most salient morphological feature of nerve cells. Indeed, there exist not only nerve cells that are connected to just one or few other neurons but also neurons connected up to 100,000 other neurons. In fact, the dendrites and the cell body are the major sites for synaptic terminals made by the axonal endings of other neurons. The axon is a unique extension from the cell body that may travel a few hundred micrometers or much farther, depending on the type of neuron. For instance, the axons running from the human spinal cord to the foot are about a meter long. The axonal mechanism that carries signals over such distances is called "action potential," a self-regenerating wave of electrical activity that propagates from the cell body to the axon terminal ending with a set of presynaptic buttons. The

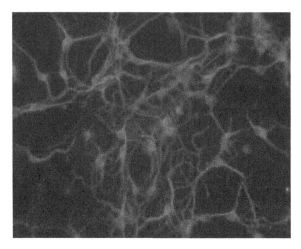

Fig. 1 Network of embryonic rat fore-brain neurons and synapses cultured in vitro. Courtesy of Carla Perrone Capano

synaptic terminal is a special secretory apparatus, where typically a presynaptic terminal is adjacent to a postsynaptic specialization of the contacted neuron, with no physical continuity. Indeed, the two terminals communicate via secretion of neurotransmitters from the presynaptic terminal that bind to receptors in the postsynaptic specialization. The secretion is triggered by the arrival of the action potential to the presynaptic terminal.

The fundamental purpose of neurons is to integrate information from other neurons. The number of inputs is therefore an important determinant of neuronal function. The integrated information is sent out through the axon to synaptic connections with other neurons. The action potential is a signal of the kind "all or none": The membrane resting potential of a neuron is generally $-70\,\mathrm{mV}$; if the integrated inputs bring the potential to a threshold value of about $-55\,\mathrm{mV}$, the action potential is triggered and, as a consequence, the membrane potential is depolarized. The action potential is a signal of about $100\,\mathrm{mV}$ in amplitude and $1-3\,\mathrm{ms}$ in duration. It travels along the axon by means of the successive openings and closings of the Na^+ and K^+ voltage gated channels, which allow transverse ion flow through the axon membrane. After the activation of ion channels, a certain time is needed to reset these channels to the resting state. This time interval is the "refractory period," during which the channels are inactive, i.e. they cannot be opened even in the presence of a stimulus over the threshold. During the refractory period the neuron is unable to answer to any stimulus, regardless its intensity, and therefore another action potential is impossible to elicit. The refractoriness of the membrane in the wake of the action potential explains why action potentials do not propagate back toward the point of their initiation as they travel along the axon.

The arrival of the action potential at the presynaptic buttons gives rise to synaptic transmission usually via chemical synapses, the most common in the nervous system (another type, electrical synapse, is less frequent). In the typical chemical synapse, the arrival of the action potential in the presynaptic terminal opens the voltage gated Ca^{2+} channels, leading to the influx of Ca^{2+} ions. These cause the vesicles containing neurotransmitter to fuse with the presynaptic membrane. The neurotransmitter is then released into the synaptic cleft and can bind to receptor molecules in the postsynaptic membrane, opening postsynaptic ion channels. Depending on the neurotransmitter (e.g., glutamate or GABA) and the receptor to which it binds, different ion channels will be opened driving the postsynaptic membrane potential closer or farther with respect to the threshold potential. Neurotransmitters can then have an excitatory or inhibitory effect (glutamate and GABA respectively) depending on the resulting variation of the postsynaptic membrane potential.

In 1949 Donald Hebb first proposed that the strength of a synaptic connection depends on the activity of the connection itself [23]. He stated his postulate: "When an axon of cell A is near enough to excite a cell B and repeatedly or persistently takes part in firing it, some growth process or metabolic change takes place in one or both cells such that A's efficiency, as one of the cell firing B, is increased." This strengthening has its counterpart in the complement postulate stating that if a neuron A repeatedly fails to excite neuron B, the strength of the connection is decreased. The pioneering work of Hebb was confirmed only about 20 years later by the recognition of the long term potentiation (LTP) and long term depression (LTD). These mechanisms can be induced in vivo by the application of high frequency (100 Hz for 1 min) for LTP, or a low frequency (1–5 Hz for 5–10 min) stimulation for LTD, producing permanent plastic modifications. LTP and LTD are generally accepted as a paradigmatic example of Hebbian plasticity on the basis of memory and learning.

3 Neural Avalanches and Networks: Experimental Results

During neuronal activity, each neuron can receive inputs by thousands of other neurons and, when it reaches a threshold, redistributes this integrated activity back to the neuronal network. Recently, a neuronal activity based on avalanches has been observed in organotypic cultures from coronal slices of rat cortex [5] where neuronal avalanches are stable for many hours [6]. More precisely, recording spontaneous local potentials continuously by a multielectrode array has shown that activity initiated at one electrode might spread to other electrodes not necessarily contiguous, as in a wave-like propagation. Cortical slices are then found to exhibit a new form of activity, producing several avalanches per hour of different duration, in which non-synchronous activity is spread over space and time. By analyzing the size and duration of neuronal avalanches, the probability distribution reveals a power law behavior, suggesting that the cortical network operates in a critical state. The experimental data indicate for the avalanche size distribution a slope varying between -1.2 and -1.9, depending on the accuracy of the time-binning procedure, with a value -1.5 for optimal experimental conditions. Interestingly, the power-law behavior is destroyed when the excitability of the system is increased, contrary to what was expected, since the incidence of large avalanches should decrease the power law exponent. The distribution then becomes bimodal, i.e. dominated by either very small or very large avalanches as in epileptic tissue. The power law behavior is therefore the indication of an optimal excitability in the system's spontaneous activity. These results have been interpreted relating spontaneous activity in a cortical network to a critical branching process [30]; indeed the experimental branching parameter is very close to the critical value equal to one, at which avalanches at all scales exist. Neuronal network simulations with global coupling and static synapses show, for the conservative case, similar values of the scaling exponent [31].

In real brain neurons are known to be able to develop an extremely high number of connections with other neurons, i.e. a single cell body may receive inputs from even a hundred thousand presynaptic neurons. One of the most fascinating questions is how an ensemble of living neurons self-organizes, developing connections to give origin to a highly complex system. The dynamics underlying this process should be driven by both the aim of realizing a well-connected network leading to efficient information transmission and the energetic cost of establishing very long connections. The morphological characterization of a neuronal network grown in vitro has been studied [32] by monitoring the development of neurites in an ensemble of few hundred neurons from the frontal ganglion of adult locusts. After few days the cultured neurons have developed an elaborated network with hundreds of connections, whose morphology and topology has been analyzed by mapping it onto a connected graph. The short path length and the high clustering coefficient measured indicate that the network belongs to the category of small-world networks [33], interpolating between regular and random networks.

4 The Model

In order to formulate a new model to study neuronal activity, we have introduced [19] within a SOC approach the three most important ingredients, namely threshold firing, neuron refractory period, and activity-dependent synaptic plasticity. We consider a simple square lattice of size $L \times L$ on which each site represents the cell body of a neuron, each bond a synapse. Therefore, on each site we have a potential v_i and on each bond a conductance g_{ij}. Whenever at time t the value of the potential at a site i is above a certain threshold $v_i \geq v_{\max}$, approximately equal to $-55\,\mathrm{mV}$ for the real brain, the neuron fires, i.e. generates an "action potential," distributing charges to its connected neighbors in proportion to the current flowing through each bond

$$v_j(t+1) = v_j(t) + v_i(t)\frac{i_{ij}(t)}{\sum_k i_{ik}(t)} \tag{1}$$

where $v_j(t)$ is the potential at time t of site j, nearest neighbor of site i, $i_{ij} = g_{ij}(v_i - v_j)$, and the sum is extended to all nearest neighbors k of site i that are at a potential $v_k < v_i$. After firing a neuron is set to a zero-resting potential. The conductances are initially all set equal to unity whereas the neuron potentials are uniformly distributed random numbers between $v_{max} - 2$ and $v_{max} - 1$. The potential is fixed to zero at top and bottom whereas periodic boundaries are imposed in the other direction.

The external stimulus is imposed either at one input site in the center of the lattice or at random in the system, this last choice more closely modeling brain spontaneous activity. The electrical activity is monitored as a function of time by measuring the total current flowing in the system. The firing rate of real neurons is limited by the refractory period, i.e. the brief period after the generation of an action potential during which a second action potential is difficult or impossible to elicit. The practical implication of refractory periods is that the action potential does not propagate back toward the initiation point and therefore is not allowed to reverberate between the cell body and the synapse. In our model, once a neuron fires, it remains quiescent for one time step and it is therefore unable to accept charge from firing neighbors. This ingredient indeed turns out to be crucial for a controlled functioning of our numerical model. In this way, an avalanche of charges can propagate far from the input through the system.

As soon as a site is at or above threshold v_{max} at a given time t, it fires according to (1), then the conductance of all the bonds, connecting to active neurons and that have carried a current, is increased in the following way

$$g_{ij}(t+1) = g_{ij}(t) + \delta g_{ij}(t) \tag{2}$$

where $\delta g_{ij}(t) = k\alpha i_{ij}(t)$, with α being a dimensionless parameter and k a unit constant bearing the dimension of an inverse potential. After applying (2) the time variable of our simulation is increased by one unit. Equation (2) describes the mechanism of increase of synaptic strength, tuned by the parameter α. This parameter then represents the ensemble of all possible physiological factors influencing synaptic plasticity, many of which are not yet fully understood.

Once an avalanche of firings comes to an end, the conductance of all the bonds with non-zero conductance is reduced by the average conductance increase per bond:

$$\Delta g = \sum_{ij,t} \delta g_{ij}(t)/N_b \tag{3}$$

where N_b is the number of bonds with non-zero conductance. The quantity Δg depends on α and on the response of the brain to a given stimulus. In this way, our electrical network "memorizes" the most used paths of discharge by increasing their conductance, whereas the less used synapses atrophy. Once the conductance of a bond is below an assigned small value σ_t, we remove it, i.e. set it equal to zero, which corresponds to what is known as pruning. This remodeling of synapses mimics the fine tuning of wiring that occurs during "critical periods" in the developing brain, when neuronal activity can modify the synaptic circuitry, once the basic patterns of brain wiring are established [21]. These mechanisms correspond to a Hebbian form of activity-dependent plasticity, where the conjunction of activity at the presynaptic and postsynaptic neuron modulates the efficiency of the synapse [26]. To insure the stable functioning of neural circuits, both strengthening and weakening rule of Hebbian synapses are necessary to avoid instabilities due to positive feed-back [41]. However, differently from the well-known LTP and LTD mechanisms, in our model the modulation of synaptic strength does not depend on the frequency of synapse activation [20, 42, 43]. It should also be considered that in the living brain synapses exhibiting plasticity are not electrical but chemical. For instance, Hebbian plasticity at excitatory synapses is classically mediated by postsynaptic calcium-dependent mechanisms [34]. In our approach, the excitability of the postsynaptic neuron is simply modulated by the value of the electrical potential.

5 Pruning and Neuronal Avalanches

The external driving mechanism to the system is imposed by setting the potential of the input site to the value v_{max}, corresponding to one stimulus. This external stimulus is needed to keep functioning the system and therefore mimics the living brain activity. We let the discharge evolve until no further firing occurs, then we apply the next stimulus. Figure 2 shows the electrical signal as a function of time: the total current flowing in the system is recorded in time during a sequence of successive avalanches. As defined above, the time unit corresponds to the time necessary to propagate the signal from a neuron to the next nearest neighbors. Data show

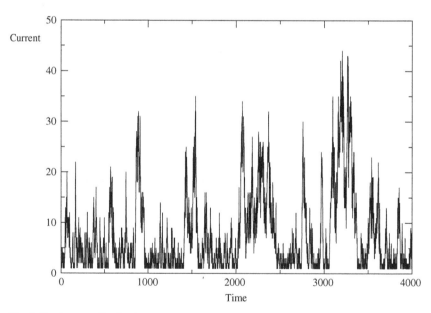

Fig. 2 Total current flowing in one square lattice configuration ($L = 1000$, $\alpha = 0.03$, $\sigma_t = 0.0001$, $v_{max} = 6$) as function of time in a sequence of several thousand stimuli

that discharges of all sizes are present in the brain response, as in self-organized criticality where the avalanche size distribution scales as a power law [5, 35][1].

The strength of the parameter α, controlling both the increase and the decrease of synaptic strength, determines the plasticity dynamics in the network. In fact, the more the system learns strengthening the used synapses , the more the unused connections will weaken. The number of pruned bonds in the system as function of time indicates that, for large values of α the system strengthens more intensively the synapses carrying current but also very rapidly prunes the less used connections, reaching after a short transient a plateau where it prunes very few bonds. On the contrary, for small values of α the system takes more time to initiate the pruning process and slowly reaches a plateau (Fig. 3). The inset of Fig. 3 shows the asymptotic value of the fraction of active bonds, calculated as the total number of bonds in the unpruned network minus the asymptotic number of pruned bonds, as function of α. The number of active (non-pruned) bonds asymptotically reaches its largest

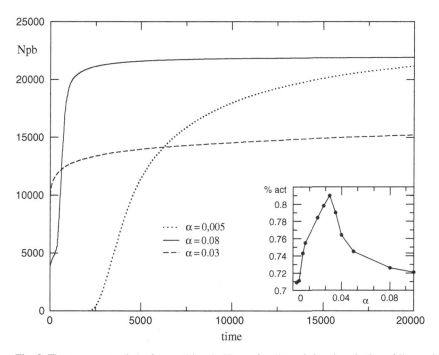

Fig. 3 The average number of pruned bonds N_{pb} as function of time in a lattice of linear size $L = 100$ for different values of α. The value of the parameters is $\sigma_t = 0.0001$ and $v_{max} = 6$. In the inset we show the asymptotic value of the percentage of active bonds as function of α

---

[1] The total current shows a pattern reminiscent of spontaneously occuring neuronal oscillations observed in developmental central nervous system networks, such as the so-called "giant depolarizing potentials," network-driven synaptic events observed in the immature hippocampus, when all circuits are excitatory.

value at the value $\alpha = 0.03$. This could be interpreted as an optimal value for the system with respect to plastic adaptation.

Since each avalanche may trigger the activity of a high number of neurons, large currents flow through the system; therefore after N_p stimuli the network is no longer a simple square lattice due to pruning, but exhibits a ladder-like pattern with few lateral connections. This complex structure constitutes the first approximation to a trained brain, on which measurements are performed. These consist of a new sequence of stimuli at the input site, by setting the voltage at threshold, during which we measure the number of firing neurons as function of time. This quantity corresponds to the total current flowing in a discharge measured by the electromagnetic signal of the EEG. We have evaluated the size distribution of neural avalanches, i.e. the total number of neurons involved in the propagation of each stimulus. This distribution exhibits power-law behavior, with an exponent equal to 1.2 ± 0.1, quite stable with respect to parameters (Fig. 4). We have also simulated the brain dynamics on a square lattice with a small fraction of bonds, from 0 to 10%, rewired to long-range connections corresponding to a small-world network [32, 33, 36], which more realistically reproduces the connections in the real brain. Figure 4 shows the size distribution scaling with an exponent 1.2 ± 0.1 for a system with 1% rewired bonds and a different set of parameters, α, N_p, and $v_{\max}$. Conversely, for the input

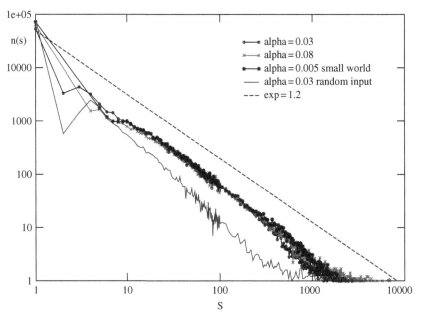

Fig. 4 Log-log plot of the distribution of avalanche size $n(s)$ ($L = 1000$, $\alpha = 0.03$ and 0.08, $N_p = 10$, $v_{\max} = 6$) for the square lattice (*lines*) and the small world lattice (*, $L = 1000$, $\alpha = 0.05$, $N_p = 1000$, $v_{\max} = 8$) with 1% rewired bonds. The data are averaged over 10,000 stimuli in 10 different configurations and the stimulus is applied at the central site. The *dashed line* has a slope 1.2. For $\alpha = 0.3$ and random input site the slope is 1.5

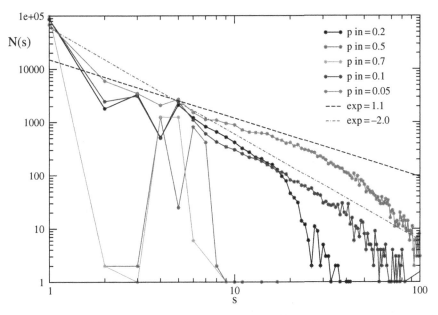

Fig. 5 Log-log plot of the distribution of avalanche size $n(s)$ ($L = 1000$, $\alpha = 0.03$, $N_p = 10$, $v_{\max} = 6$) for the square lattice and different probability of inhibitory synapses p_{in}. The data are averaged over 10,000 stimuli in 10 different configurations. For $p_{\text{in}} > 20\%$ the scaling behavior is lost

site chosen at random in the system, the scaling exponent changes and becomes 1.5 ± 0.1 (Fig. 4) in agreement with the experimental value found for spontaneous activity [5].

In the mature living brain, synapses can be excitatory or inhibitory, namely they set the potential of the postsynaptic membrane to a level closer or farther, respectively, to the firing threshold. We have introduced in our model this ingredient: each synapse is inhibitory with probability p_{in} and excitatory with probability $1 - p_{\text{in}}$. The size distribution (Fig. 5) exhibits power-law behavior for densities of inhibitory synapses up to 10%, whereas the scaling behavior is lost for higher densities.

6 Power Spectra

In order to compare with medical data, we calculate the power spectrum of the resulting time series, i.e. the square of the amplitude of the Fourier transform as function of frequency. The average power spectrum as function of frequency is shown in a log-log-plot with the parameters $\alpha = 0.03$, $N_p = 10$, $\sigma_t = 0.0001$, $v_{\max} = 6$ and a lattice of size $L = 1000$ (Fig. 6). We see that it exhibits a power-law behavior with the exponent 0.8 ± 0.1 over more than three orders of magnitude. This is the same value for the exponent found generically on medical EEG power spectra [37, 38].

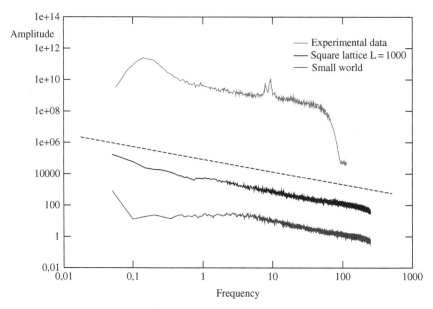

Fig. 6 Power spectra for experimental data and numerical data ($L = 1000$, $\alpha = 0.03$, $N_p = 10$, $v_{\max} = 6$) for the square lattice (*middle line*) and the small world lattice (*bottom line*, $L = 1000$, $\alpha = 0.05$, $N_p = 1000$, $v_{\max} = 8$) with 1% rewired bonds. The experimental data (*top line*) are from [38] and frequency is in Hz. The numerical data are averaged over 10,000 stimuli in 10 different network configurations. The *dashed line* has a slope 0.8

We also show in Fig. 6 the magnetoelectroencephalography (similar to EEG) obtained from channel 17 in the left hemisphere of a male subject resting with his eyes closed, as measured in [38], having the exponent 0.795.

We have checked that the value of the exponent is stable against changes of the parameters α, $v_{\max}$, σ_t, and N_p, and also for random initial bond conductances. Moreover, the scaling behavior remains unchanged if the input site is placed at random in the system at each stimulus. For $\alpha = 0$ the frequency range of validity of the power law decreases by more than an order of magnitude. Figure 6 also shows the power spectrum for a small-world network with 1% rewired bonds and a different set of parameters, α, N_p, and $v_{\max}$: the spectrum has some deviations from the power law at small frequencies and tends to the same universal scaling behavior at larger frequencies over two orders of magnitude. The same behavior is found for a larger fraction of rewired bonds.

The scaling behavior of the power spectrum can be interpreted in terms of a stochastic process determined by multiple random inputs [39]. In fact, the output signal resulting from different and uncorrelated superimposed processes is characterized by a power spectrum with power-law behavior and a crossover to white noise at low frequencies. The crossover frequency is related to the inverse of the longest characteristic time among the superimposed processes. The value of the scaling exponent depends on the ratio of the relative effect of a process of given frequency on the output with respect to other processes. $1/f$ noise corresponds to a superposition of processes of different frequency having all the same relative effect on the output signal. In our case the scaling exponent is smaller than unity, suggesting that processes with high characteristic frequency are more relevant than processes with low frequency in the superposition [39].

In real systems, neurons may have a leakage, namely the potential decays exponentially in time with a relaxation time τ, i.e. $\frac{dv(t)}{dt} = -\gamma v(t)$, with $\gamma = 1/\tau$. Leakage has been considered in our model and the same scaling behavior has been recovered (Fig. 7). However, for $\tau \leq 10$ (i.e., for stronger leaking), the low frequency part of the spectrum appears to be frequency independent and the scaling regime is recovered at high frequencies with an exponent in agreement with the previous results.

We have also studied the power spectrum for a range of value of p_{in}. For a density up to 10% of inhibitory synapses, the same power-law behavior is recovered within error bars (Fig. 8). For increasing density, the scaling behavior is progressively lost and the spectrum develops a complex multi-peak structure for $p_{\text{in}} = 0.5$. These results suggest that the balance between excitatory and inhibitory synapses has a crucial role on the overall behavior of the network, similarly to what can occur in some severe neurological and psychiatric disorders [40].

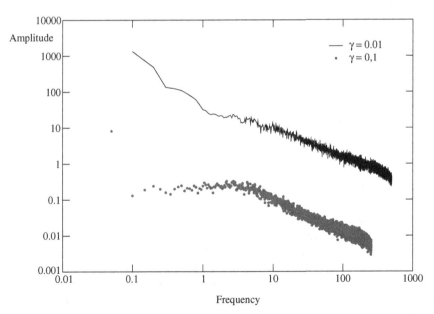

Fig. 7 Power spectra for numerical data (square lattice $L = 1000$, $\alpha = 0.03$, $N_p = 10$, $v_{\text{max}} = 6$) for the case of leaky neurons with two different values of γ. Data are averaged over 10,000 stimuli at the central site in 10 different network configurations

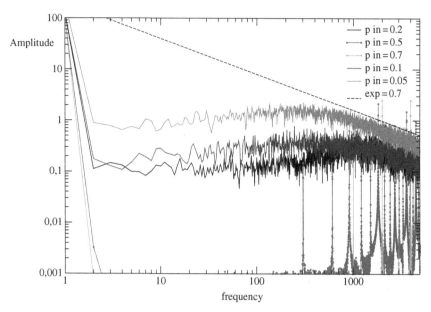

Fig. 8 Power spectra for numerical data (square lattice $L = 1000$, $\alpha = 0.03$, $N_p = 10$, $v_{max} = 6$) for different values of the probability of inhibitory synapses p_{in}. Data are averaged over 10,000 stimuli at the central site in 10 different network configurations

7 Conclusions

Extensive simulations have been presented for a novel activity-dependent brain model implemented on regular and small-world lattices. The results are compared with experimental data. An interesting result is that an optimal value of the plasticity strength α exists with respect to the pruning process that optimizes information transmission. Moreover, the avalanche size distribution shows a power-law behavior with an exponent 1.5 ± 0.1 for random input site. This value is compatible with 1.5 ± 0.4, experimentally found for neuronal avalanches.

The stability of the spectrum exponent suggests that an universal scaling characterizes a large class of brain models and physiological signal spectra for brain-controlled activities. Medical studies of EEG focus on subtle details of a power spectrum (e.g., shift in peaks) to discern between various pathologies. These detailed structures however live on a background power-law spectrum that shows universally an exponent of about 0.8, as measured for instance in [37] and [38]. A similar exponent was also detected in the spectral analysis of the stride-to-stride fluctuations in the normal human gait which can directly be related to neurological activity [16]. The discussed model is based on SOC ideas: the threshold dynamics insures time-scale separation (slow external drive and fast internal relaxation). This dynamics leads to criticality and therefore power-law behavior [2]. However, the new ingredients of the model, namely the plasticity of the synapses, may be at the origin of the new observed exponent. This work may open new perspectives to study pathological features of EEG spectra by including further realistic details into the neuron and synapses behavior.

Acknowledgements We gratefully thank E. Novikov and collaborators for allowing us to use their experimental data. We also thank Salvatore Striano, MD, for discussions and Stefan Nielsen and Hansjörg Seybold for help. L.d.A. research is supported by EU Network MRTN-CT-2003-504712, MIUR-PRIN 2004, MIUR-FIRB 2001, CRdC-AMRA, and INFM-PCI.

References

1. P. Bak: *How nature works. The Science of Self-Organized Criticality*, Springer, New York, 1996
2. H.J. Jensen: *Self-Organized Criticality*, Cambridge University Press, Cambridge, 1998
3. S. Maslov, M. Paczuski, P. Bak: Phys. Rev. Lett. **73**, 2162 (1994)
4. J. Davidsen, M. Paczuski: Phys. Rev. E **66**, 050101(R) (2002)
5. J.M. Beggs, D. Plenz: J. Neurosci. **23**, 11167 (2003)
6. J.M. Beggs, D. Plenz: J. Neurosci. **24**, 5216 (2004)
7. P. Bak, C. Tang: J. Geophys. Res. **94**, 15635 (1989)
8. A. Sornette, D.Sornette: Europhys. Lett. **9**, 197 (1989)
9. P. Bak, K. Sneppen: Phys. Rev. Lett. **71**, 4083 (1993)
10. E.T. Lu, R.J. Hamilton: Astrophys. J. **380**, L89 (1991)
11. P.A. Politzer: Phys. Rev. Lett. **84**, 1192 (2000)
12. J. Faillettaz, F. Louchet, J.R. Grasso: Phys. Rev. Lett. **93**, 208001 (2004)
13. O. Peters, C. Hertlein and K. Christensen: Phys. Rev. Lett. **88**, 018701 (2002)
14. A. Gevins et al.: Trends Neurosci. **18**, 429 (1995)
15. G. Buzsaki, A. Draguhn: Science **304**, 1926 (2004)
16. J.M. Hausdorff et al.: Physica A. **302**, 138 (2001)
17. S.B. Lowen, S.S. Cash, M. Poo, M.C. Teich: J. Neurosci. **17**, 5666 (1997)
18. D.R. Chialvo, G.A. Cecchi, M.O. Magnasco: Phys. Rev. E **61**, 5654 (2000)
19. L. de Arcangelis, C. Perrone-Capano, H.J. Herrmann: Phys. Rev. Lett. **96**, 028107 (2006)
20. T.D. Albraight et al.: Neuron, Review supplement to vol. **59** (February 2000)
21. T.K. Hensch: Ann. Rev. Neurosci. **27**, 549 (2004)
22. L.F. Abbott, S.B. Nelson: Nature Neurosci. **3**, 1178 (2000)
23. D.O. Hebb: *The Organization of Behaviour*, John Wiley, New York, 1949
24. J.Z. Tsien: Curr. Opin. Neurobiol. **10**, 266 (2000)
25. G.-Q. Bi, M.-M. Poo: Ann. Rev. Neurosci. **24**, 139 (2001)
26. S.J. Cooper: Neurosci. Biobehav. Rev. **28**, 851 (2005)
27. Y. Ashkenazy et al.: Physica A. **316**, 662 (2002)
28. P. Ch. Ivanov et al.: Europhys. Letters **43**, 363 (1998)
29. D. Purves, G.J. Augustine, D. Fitzpatrick, W.C. Hall: *Neuroscience*, Chap. 1–6, Sinauer, Sunderland 2004
30. S. Zapperi, K.B. Lauritsen, H.E. Stanley: Phys. Rev. Lett. **75**, 4071 (1995)
31. C.W. Eurich, J.M. Herrmann, U.A. Ernst: Phys. Rev. E **66**, 066137 (2002)
32. O. Shefi, I. Golding, R. Segev, E. Ben-Jacob, A. Ayali: Phys. Rev. E. **66**, 021905 (2002)
33. D.J. Watts, S.H. Strogatz: Nature **393**, 440 (1998)
34. G.Q. Bi, M.M. Poo: Annu. Rev. Neurosci. **24**, 139 (2001)
35. A.M. Kasyanov, V.F. Safiulina, L.L. Voronin, E. Cherubini: Proc Natl Acad Sci USA, 1013967–3972 (2004)
36. L.F. Lago-Fernandez, R. Huerta, F. Corbacho, J.A. Siguenza: Phys. Rev. Lett. **84**, 2758 (2000)

37. W.J. Freeman et al.: J. Neurosci. Methods **95**, 111 (2000)
38. E. Novikov, A. Novikov, D. Shannahoff-Khalsa, B. Schwartz, J. Wright: Physical Rev. E. **56**, R2387 (1997)
39. J.M. Hausdorff, C.K. Peng: Phys. Rev. E. **54**, 2154 (1996)
40. J.L. Rubenstein, M.M. Merzenich: Genes Brain Behav. **2**, 255 (2003); E.M. Powell et al.: J. Neurosci. **2**, 622 (2003)
41. N.S. Desai: J. Physiol. Paris **97**, 391 (2003)
42. O. Paulsen, T.J. Sejnowski: Curr. Opin. Neurobiol. **10**, 172 (2000)
43. K.H. Braunewell, D. Manahan-Vaughan: Rev. Neurosci. **12**, 121 (2001)

Index